VOLUME ONE HUNDRED AND TWENTY

ADVANCES IN
COMPUTERS

AI and Cloud Computing

VOLUME ONE HUNDRED AND TWENTY

ADVANCES IN
COMPUTERS

AI and Cloud Computing

Edited by

ALI R. HURSON
*Missouri University of Science and Technology,
Rolla, MO, United States*

SHENG WU
*UN/UN Agency,
Geneva, Switzerland*

ACADEMIC PRESS

An imprint of Elsevier

ELSEVIER

Academic Press is an imprint of Elsevier
50 Hampshire Street, 5th Floor, Cambridge, MA 02139, United States
525 B Street, Suite 1650, San Diego, CA 92101, United States
The Boulevard, Langford Lane, Kidlington, Oxford OX5 1GB, United Kingdom
125 London Wall, London, EC2Y 5AS, United Kingdom

First edition 2021

ISBN: 978-0-12-821147-2
ISSN: 0065-2458

For information on all Academic Press publications
visit our website at https://www.elsevier.com/books-and-journals

Publisher: Zoe Kruze
Editorial Project Manager: Leticia Lima
Production Project Manager: James Selvam
Cover Designer: Alan Studholme

Typeset by SPi Global, India

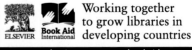

Working together
to grow libraries in
developing countries

www.elsevier.com • www.bookaid.org

Contents

6. Protecting personal sensitive data security in the cloud with blockchain   **195**

Zhen Yang, Yingying Chen, Yongfeng Huang, and Xing Li

Contributors

Enguo Cao
School of Design, Jiangnan University, Wuxi, Jiangsu Province, PR China

Yingying Chen
School of Engineering and Applied Science, University of Virginia, Charlottesville, VA, United States

Qi Cui
School of Computer and Software, Nanjing University of Information Science and Technology, Nanjing, China

Honglei Guo
School of Design, Jiangnan University, Wuxi, Jiangsu Province, PR China

Yongfeng Huang
Department of Electronic Engineering, Tsinghua University, Beijing National Research Center for Information Science and Technology, Beijing, China

Leqi Jiang
School of Computer and Software; Jiangsu Engineering Center of Network Monitoring, Nanjing University of Information Science and Technology, Nanjing, China

Xing Li
Department of Electronic Engineering, Tsinghua University, Beijing National Research Center for Information Science and Technology, Beijing, China

Huanhuan Lian
Henan Polytechnic University, Jiaozuo, China

Xingming Sun
School of Computer and Software; Jiangsu Engineering Center of Network Monitoring, Nanjing University of Information Science and Technology, Nanjing, China

Yongli Tang
Henan Polytechnic University, Jiaozuo, China

Lizhi Wang
School of Computer Science and Technology, Donghua University, Shanghai, PR China

Pengwei Wang
School of Computer Science and Technology, Donghua University, Shanghai, PR China

Xiaojun Wang
Dublin City University, Dublin, Ireland

Q.M. Jonathan Wu
Department of Electrical and Computer Engineering, University of Windsor, Windsor, ON, Canada

Sheng Wu
Chinese Academy of Sciences, Beijing, China

Zhihua Xia
School of Computer and Software; Jiangsu Engineering Center of Network Monitoring, Nanjing University of Information Science and Technology, Nanjing, China

Bin Yang
School of Design, Jiangnan University, Wuxi, Jiangsu Province, PR China

Zhen Yang
Department of Electronic Engineering, Tsinghua University, Beijing National Research Center for Information Science and Technology, Beijing, China

Jinxia Yu
Henan Polytechnic University, Jiaozuo, China

Chengsheng Yuan
School of Computer and Software, Nanjing University of Information Science and Technology, Nanjing, China; Department of Electrical and Computer Engineering, University of Windsor, Windsor, ON, Canada

Xiaobo Zhang
School of Computer Science and Technology, Donghua University, Shanghai, PR China

Zhaohui Zhang
School of Computer Science and Technology, Donghua University, Shanghai, PR China

Zongqu Zhao
Henan Polytechnic University, Jiaozuo, China

Yongjun Zheng
School of Physics, Engineering and Computer Science, University of Hertfordshire, Hertfordshire, United Kingdom

Xinxin Zhou
School of Computer Science and Technology, Donghua University, Shanghai, PR China

Preface

Traditionally, *Advances in Computers*, the oldest series to chronicle of the rapid evolution of computing, annually publishers several volumes, each one typically comprised of four to eight chapters, describing new developments in the theory and applications of computing.

The 120th volume is an eclectic volume inspired by recent advances in artificial intelligence (AI) and cloud computing, and their interactions. AI and cloud computing jointly drive the Fourth Industrial Revolution. A blend of these two technologies has resulted in unprecedented advances in computing power, algorithm, theory, and application. While most of researchers and practitioners have been paying attention to how AI works in cloud computing environment. This volume focuses on how AI can work for cloud computing environment, resolving the pressing challenges in the cloud era, including cybersecurity, privacy protection, and anticounterfeit.

This volume is a collection of six chapters and each chapter explores innovative AI solutions to tackle major challenges in cloud environment:

In Chapter 1, "A deep-forest based approach for detecting fraudulent online transaction," by Zhang et al. articulates fundamentals of AI's application in fraud detection and analyzes the key design considerations for online transaction fraud detection. Through the case study of one machine learning algorithm, the authors examine the ways the challenges in detecting online transaction fraud, i.e., the problems of big computing load and unbalanced distribution, are met. Finally, a deep forest-based approach for online transactions fraud detection is proposed and evaluated based on the detection precision rate metric.

Chapter 2 entitled "Design of cyber-physical-social systems with forensic-awareness based on deep learning" by Yang et al. proposes a scheme of forensics-awareness in support of reliable image forensics investigations. The proposed forensic scheme utilizes a multichannel convolutional neural network (MCNN) to learn hierarchical representations from the input images. Unlike most existing forensic methods, which can only detect a certain type of image tampering, this system can capture various forgery operations, including clone, removal, splicing, and smoothing.

To enhance cloud computing security, protocols reviews and enhancement are of significance in both data privacy and password protection. Usually two major steps are involved in data operation: encryption and construction. Chapter 3, "Review on privacy-preserving data comparison protocols in cloud computing," by Jiang et al. focuses on the second step, where research remain scattered. The chapter presents an in-depth review of the development of the privacy-preserving data comparison. Several most commonly used privacy-preserving data comparison protocols are introduced, analyzed, and their applications based on metrics such as functionality, security, computation complexity, and communication overhead are articulated.

Chapter 4, "Provably secure verifier-based password authenticated key exchange based on lattice," by Yu et al. proposes a new VPAKE protocol based on lattices, which is constructed by using chosen ciphertext attacks (CCA) secure public key encryption scheme. The protocol is robust against quantum attacks. This scheme enables ASCII-based passwords and a zero-knowledge password policy check. In addition, it allows users to prove the compliance of their password without revealing any information.

AI offers optimized solutions for existing challenges, and the cloud environment conditions keep motivating researchers to explore new solutions that meet application demands. Chapter 5, "Fingerprint liveness detection using an improved CNN with the spatial pyramid pooling structure," by Yuan et al. proposes a novel fingerprint liveness detection (FLD) method based on an improved convolutional neural network (CNN) with spatial pyramid pooling (ICNNSPP) to overcome the limitation of input image size/scale, which is one of the bottlenecks of most exiting CNN for FLD.

In many application domains, it is a valid assumption that the data owner is concern of privacy and security of the personal information. Meeting this requirement is challenging when the data is outsourced to the cloud. Chapter 6, "Protecting personal sensitive data security in the cloud with blockchain," by Yang et al. introduces a trust-free data access model for personal sensitive data protection in the cloud environment. In this model, an access control mechanism is constructed based on the Ethereum blockchain, which requires no trusted party. The smart contract enables fine-grained access control for cloud data based on the blockchain. Data operations including uploading, updating, and downloading can be automatically processed and logged to ensure transparency and auditability.

We hope that readers find this volume of interest, and useful for teaching, research, and other professional activities. We welcome feedback on the volume, as well as suggestions for topics of future volumes. Finally, we would like to thank Dr. Xianyi Chen and Dr. Zhihua Xia for their efforts in facilitating this volume.

ALI R. HURSON
Missouri University of Science and Technology,
Rolla, MO, United States
SHENG WU
UN/UN Agency, Geneva, Switzerland

CHAPTER ONE

A Deep-forest based approach for detecting fraudulent online transaction

Lizhi Wang[a], Zhaohui Zhang[a], Xiaobo Zhang[a], Xinxin Zhou[a], Pengwei Wang[a], and Yongjun Zheng[b]

[a]School of Computer Science and Technology, Donghua University, Shanghai, PR China
[b]School of Physics, Engineering and Computer Science, University of Hertfordshire, Hertfordshire, United Kingdom

Contents

Advances in Computers, Volume 120
ISSN 0065-2458
https://doi.org/10.1016/bs.adcom.2020.10.001

1

Abstract

Fraudulent transaction is the one of the most serious threats to online security nowadays. Artificial Intelligence is vital for financial risk control in cloud environment. Many studies had attempted to explore methods for online transaction fraud detection; however, the existing methods are not sufficient to conduction detection with high precision. In this chapter, we propose a Deep-forest based approach for online transactions fraud detection, which integrates differentiation feature generation method and deep-forest based model. As a single-time transaction's information, which does not contain information such as the user's behavior, is not sufficient for detecting fraudulent transaction, we introduce a transaction time-based differentiation feature generation method into our scheme. *Individual Credibility Degree (ICD)* and *Group Anomaly Degree (GAD)*, which are based on transaction time, are derived to distinguish between legal and fraudulent transaction. Furthermore, to deal with the extreme imbalance of online transactions, we apply Deep-forest algorithm to detect fraudulent transactions. While raw deep-forest model could ignore the outlier transaction samples, we enhance the raw Deep forest with detection mechanism for outliers, paying more attention on outliers to promote the precision of fraud detection model. Finally, we conduct test using one bank's transaction data. Compared with random Forest-detection model, our method improves precision rate by 15% and recall rate by 20%.

Abbreviations

AI	Artificial Intelligence
AVS	Address Verification Systems
B2C	Business–to–Customer
CART	Classification and Regression Tree
CNN	convolutional Neural Networks
CVM	Card Verification Method
DBN	Deep Belief Networks
DNN	deep neural networks
FN	False Negative
FP	False Positive
GAD	Group Anomaly Degree
gcForest	multi-Grained Cascade Forest, Deep Forest
ICD	Individual Credibility Degree
ML	Machine Learning
PIN	Personal Identification Number
RNN	Recurrent Neural Networks
SMOTE	Synthetic Minority Oversampling Technique
WOE	Weight of Evidence
XGBoost	eXtreme Gradient Boosting

1. Introduction

Online transaction is widely used nowadays, and online transaction fraud is becoming a serious threat to financial security. When online transaction fraud occurs, the banks will suffer from financial losses and credibility damage. To individual clients, it not only means loss of money, but also brings them psychological shadow. However, the harm of online transaction fraud is difficult to measure. On the one hand, many banks and companies are unwilling to disclose the cases, and on the other hand, the delay of time in fraud detection hinders the assessment of losses.

Online transaction fraud can take place in many ways [1], including stolen card fraud, cardholder-not-present fraud, application fraud, etc. Stolen card fraud is the most common type of fraud where the fraudster usually tries to spend as soon as possible; Cardholder-not-present fraud is often observed in e-business and application fraud corresponds to the application for a card with false personal information [2]. It is worth noting that with quick development of technology, the battle between fraudsters and fraud detection is becoming more and more intense. For instance, online fraud has become China's third largest black industrial chain [3]. Researchers need to improve fraud detection methods quick enough to keep them precision and valid. In recent years, Artificial Intelligence (AI) has been used in online fraud detection efforts. By employing machine learning technology, researchers have greatly improved security solutions for online transactions.

There are two ways to combat fraudulent online transaction [2], fraud prevention and fraud detection. Fraud prevention attempts to prevent the online transaction fraud at source. Fraud prevention methods include Address Verification Systems (AVS), Card Verification Method (CVM) and Personal Identification Number (PIN). AVS verifies the address with zip code of the customer; CVM and PIN check the numeric codes input by the customers. Fraud detection detects online transaction fraud real-time. Transactions are filtered by certain rules and examined by various methods, such as mining methods.

Fraud detection is realized by two sets of models: Expert driven models and Data driven models. Expert driven models use domain knowledge from fraud investigators to define rules that are used to predict the probability of a new transaction to be fraudulent, which require manual tuning and human supervision. Data driven models set up a fraud detection system based on Machine Learning (ML) [4] able to learn from data in a supervised or unsupervised manner which patterns are the most probably related to a fraudulent behavior. It is the best solution to combine Data driven models and Expert driven models. Typically, Expert driven models are mostly used to make sure that all frauds are detected at the cost of having few false alerts, while Data driven models are used to obtain precise alerts. Expert driven solutions are the traditional methods for fraud detection, Data driven solutions are as a support to obtain a more accurate detection, but rather than as a tool to replace Expert driven models.

Effectiveness and precision are two important indicators in fraud detection. Furthermore, cost of the fraud detection process is another major concern in research. The cost of fraud detection increases with the online transaction amount [5]. Screening 2% of transactions can reduce losses caused by fraud to 1% of the total transaction value. While examining 30% of transactions can significantly reduce fraud losses caused by fraud to 0.06%, it will significantly increase the cost. In order to minimize the cost of detection, it is important to make a first screen using Expert driven model and Data driven model. Afterwards, fraud investigators could pay attention to the transactions with high risk.

The transaction process in Fig. 1 usually includes the following steps:
➢ Check the input of forgery detection model, and select transaction samples for model training. In the meanwhile, identify problems in the sample, such as problem of unbalanced data.
➢ Try some statistical based models, evaluate their performance with transaction sample and select best model.
➢ When an online transaction occurs, a fraud detection model will score the transaction. According to the score, whether this transaction is high risk or low risk of fraud will be determined. An alert threshold is set.
➢ The investigator checks the suspicious transactions, and labels the fraud transactions. At this moment, the fraud detection model will receive feedback to adjust itself for improved precision.

Traditional techniques for detecting fraudulent transactions, including Expert driven system, has many disadvantages in dealing with the complex online environment nowadays. Data driven models based on ML has drawn more and more attention from researchers and many of them have achieved good performance [6–11], but there are still many challenges ahead:

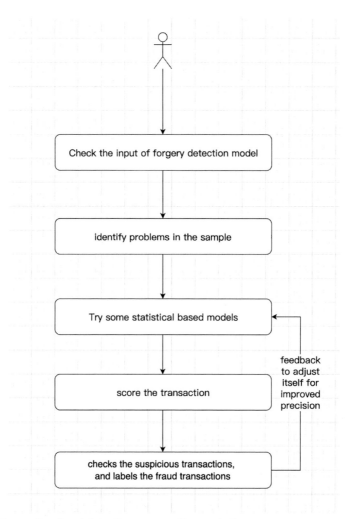

Fig. 1 Transaction fraud detection process. *Source: Author*

One major challenge is that fraud concealing techniques effect the results delivered by fraud detection model based on ML. ML-based detection models derive some features of customers' behavior pattern using a transaction aggregation technique [12], however, some fraudsters learn normal behavior patterns to bypass detection through statistical methods. Therefore, aggregation techniques alone are not sufficient to precisely distinguish legal transactions and fraudulent transactions.

Another challenge is what known as the *unbalanced problem* [13] as define by the distribution of the transactions is skewed toward the genuine class.

Since the rate of fraud transactions is usually small, machine learning algorithms are prone to neglect learning about fraudulent transactions pattern, because machine learning algorithms are constant to improve accuracy by reducing errors and do not consider the distribution of classes. Standard classification algorithms tend to focus on the class of majority, while class of minority is considered as noise.

This chapter applies one of machine learning algorithm to design a new scheme for online transaction fraud detection. The remainder of this chapter is organized as follows: Section 2 introduces Deep-forest concept we employ in our new model and its structure. Section 3 demonstrates use of machine learning in detecting online transaction fraud. Section 4 identifies the pressing issues in detecting fraud online transaction. Section 5 describes a use case for fraud detection, where deep-forest algorithm can be leveraged to resolve some major challenges in fraud detection. Section 6 summarizes key lessons learned; Section 7 concludes this chapter and provides an outlook on future research directions of machine learning in detecting fraud online transaction.

2. What is deep forest

GcForest [14] (multi-Grained Cascade Forest, Deep Forest), is a novel decision tree ensemble machine learning method. This method contains a deep-forest ensemble, with a cascade structure, which enables it conducing representation learning. Its representation learning ability can be further enhanced by multi-grained scanning when the inputs are with high dimensionality, potentially enabling gcForest to be contextual or structural aware. The number of cascade levels can be adaptively determined such that the model complexity can be automatically set, enabling gcForest to perform excellently even on small-scale data. Moreover, users can control training costs according to computational resources available.

"Go deep" is a must for learning models to tackle complicated problems. The most commonly used deep learning model is Neural Networks, a multiple layer nonlinear modules that can be trained by backpropagation. Explorations of new models that can "go deep" are on their way. Comparing with deep neural networks (DNN), another popular ML method, the gcForest has much fewer hyper-parameters and its performance is robust to hyper-parameter settings. In most cases, it can process data from different domains and get excellent results by using the default setting. Both training of gcForest and theoretical analysis with this method, is more convenient than that of DNN. In experiments in our study, gcForest achieves highly competitive performance compared with DNN, with smaller training time cost.

Our study attempts to open a door toward an alternative model, gcForest, in ML application, and to reveal its potential in online transaction fraud detection.

2.1 Basics of deep forest

2.1.1 Decision tree

Decision tree [15] is a basic classification and regression method. Decision tree consists of nodes and directed edges. There are two types of nodes: internal node and leaf node. An internal node represents a feature or attribute, and a leaf node represents a class. Fig. 2 is a diagram of a decision tree.

Decision tree model can be considered as a set of "if-then" rules; it can also be considered as conditional probability distribution defined in the feature space and the class space. Decision tree learning includes three steps: feature selection, decision tree generation and decision tree pruning.

Assume given a training set:

$$D = \{(x_1, y_1), (x_2, y_2), ..., (x_N, y_N)\} \tag{1}$$

where $x_i = (x_i^{(1)}, x_i^{(2)}, ..., x_i^{(n)})^T$ is the input, n is the number of features, $y_i = \{1, 2, ..., K\}$ is the label, $i = 1, 2, ..., N$ is the sample size. The goal of learning process is to build a decision tree model based on a given training dataset, so that it can classify the instances correctly. The algorithms commonly used in decision tree learning are ID3, C4.5 and Classification and Regression Tree (CART).

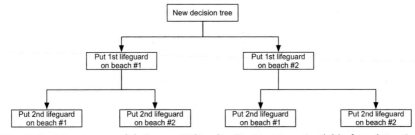

Fig. 2 Decision tree model. *Source: Wikipedia, Decision tree, Available from https:/en. wikipedia.org/wiki/Decision_tree, 1975.*

➢ In first step, *feature selection* selects classification features for the training set. It improves efficiency of the decision tree. The common criteria for feature selection are information gain, information gain ratio or Gini index.

The ID3 algorithm applies information gain criteria to select features at various nodes in the decision tree. The information gain is defined as follows:

$$g(D, A) = H(D) - H(D|A) \tag{2}$$

where D is training set, A is the feature, $H(D)$ is experience entropy of training set, $H(D|A)$ is experience entropy of feature A upon D.

The C4.5 algorithm is similar to ID3 algorithm. C4.5 improves ID3 by selecting the information gain ratio instead. The information gain ratio is defined as follows:

$$g_R(D, A) = \frac{g(D, A)}{H(D)} \tag{3}$$

where D is training set, A is the feature, $H(D)$ is experience entropy of training set, $g(D, A)$ is information gain of feature A upon D.

The CART algorithm is a widely used decision tree method, which uses Gini index to select the best feature. For a given sample set D, its Gini index is:

$$Gini(D) = 1 - \sum_{k=1}^{K} \left(\frac{|C_k|}{|D|} \right)^2 \tag{4}$$

Here, C_k is k-type sample subset in D, K is the number of classes.

If the sample set D is divided into two parts D_1 and D_2, according to whether feature A takes a certain possible value a, that is:

$$D_1 = \{(x, y) \in D | A(x) = a\}, D_2 = D - D_1 \tag{5}$$

On the condition of feature A, the Gini index of set D is defined as:

$$Gini(D, A) = \frac{|D_1|}{|D|} Gini(D_1) + \frac{|D_2|}{|D|} Gini(D_2) \tag{6}$$

➢ In the second step, the algorithm generates decision tree from root node, until all training datasets being correctly classified.

➢ Last but not the least, the final decision tree may have a good classification ability for the training data, but for the unknown test data may not have a good classification ability, that is, the fitting phenomenon may occur. The solution to this problem is to prune the decision tree that has been generated by considering the complexity of the decision tree. Decision tree pruning often cuts some leaf nodes or sub-trees above the leaf nodes, and then takes the parent node or the root node as a new leaf node. In this way, the generated decision tree is simplified.

Through the above series of operations, we aim to establish a decision tree that fits well with the training set data and has a low complexity. As directly selecting the optimal decision tree from the possible decision trees is a NP-complete problem (nondeterministic polynomial complete problem), in practice we use self-organizing method to get the suboptimal decision tree.

2.1.2 Random Forest

Random Forest [16,17] is an ensemble learning method for classification, regression and other tasks. It constructs a decision trees at training time and outputs the classes (classification). Random forest can resolve decision trees' problem in overfitting to their training set.

The *bagging method* is at the core of Random Forest. The *bagging method* consists of two parts: Bootstrap and Aggregating. Bootstrap is a random sampling process, generating dataset by simulating the true distribution of random variables. Bootstrap follows the below steps:

1. A sample is randomly taken from the dataset X;
2. Put a copy of the sample into the dataset x_j;
3. Put the sample back in X.

The above three steps are repeated N times so that there are N samples in x_j. The probability that one sample is not taken in N samples is:

$$\left(1 - \frac{1}{N}\right)^N, \text{ and } \lim_{N \to \infty} \left(1 - \frac{1}{N}\right)^N \to \frac{1}{e} \approx 0.368 \tag{7}$$

Below are steps of bagging:

1. Bootstrap generates M datasets;
2. Train M weak classifiers with M datasets;
3. The final model is a simple combination of these M weak classifiers.

Then a simple vote mechanism called Aggregating is used for classification. Taken into consideration binary classification, it is assumed that each weak classifier error rate is ϵ:

$$p(g_i(x) \neq f(x)) = \epsilon \tag{8}$$

where f is real mapping of real space in sample space, $\{G_1, G_2, ..., G_M\}$ is M weak classifier models.

Assuming that the error rate of the weak classifiers is independent of each other, it can be seen from the Hoeffding's Inequality:

$$p(G(x) \neq f(x)) = \sum_{j=0}^{\left[\frac{M}{2}\right]} \binom{M}{j} (1 - \epsilon)^j \epsilon^{M-j} \leq exp\left(-\frac{1}{2}M(1 - 2\epsilon)^2\right) \quad (9)$$

The error rate of the final model will increase exponentially as the number of weak classifiers M increases.

Random forest is a bagging algorithm with the weak classifier as a decision tree. The random forest algorithm not only samples bootstrap sampling, but also randomly selects an optional subspace of the feature space as an optional feature space of the decision tree when generating the algorithm for each node. Some Random forest classifier does not prune, but being maintained in its original form.

2.2 Cascade forest structure

While representation learning in DNN processes raw features layer-by-layer, gcForest employs a cascade structure, as illustrated in Fig. 3, where each level of cascade receives feature information processed by its preceding level, and outputs its processed result to the next level.

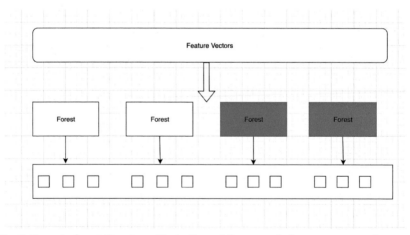

Fig. 3 Cascade forest Structure: illustration of the cascade forest structure. Suppose each level of the cascade consists of two random forests (black) and two completely random tree forests (blue). Suppose there are three classes to predict; thus, each forest will output a three-dimensional class vector, which is then concatenated for re-representation of the original input. *Source: Author*

Each level is an ensemble forests of decision tree, i.e., an ensemble of ensembles. Different types of forests are included to encourage diversity, as it is well known that diversity is crucial for ensemble construction. For simplicity, suppose that we use two completely random tree forests [17] and two random forests [18]. Each completely random tree forest contains 500 completely random trees, generated by randomly selecting a feature for split at each node of the tree, and growing tree until each leaf node contains only the same class of instances. Similarly, each random forest contains 500 trees, by randomly selecting $\sqrt{d}$ number of features as candidate (d is the number of input features) and choosing the one with the best Gini value for split.

Each forest will produce an estimate of class distribution, by counting the percentage of different classes of training examples at the leaf node where the concerned instance falls, and then averaging across all trees in the same forest, as illustrated in Fig. 4, where red color highlights paths along which the instance traverses to leaf nodes.

The estimated class distribution forms a class vector, which is then concatenated with the original feature vector to be input to the next level

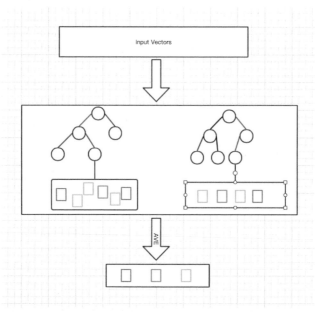

Fig. 4 Class vector generation: different marks in leaf nodes imply different classes. *Source: Author*

of cascade. For example, suppose there are three classes, then each of the four forests will produce a three-dimensional class vector; thus, the next level of cascade will receive 12 (=3 × 4) augmented features.

To reduce the risk of overfitting, class vector produced by each forest is generated by k-fold cross validation. In detail, each instance will be used as training data for $k-1$ times, resulting in $k-1$ class vectors, which are then averaged to produce the final class vector as augmented features for the next level of cascade. After expanding a new level, the performance of the whole cascade will be estimated on validation set, and the training procedure will terminate if there is no significant performance gain; thus, the number of cascade levels is automatically determined. In contrast to most deep neural networks whose model complexity is fixed, gcForest adaptively decides its model complexity by terminating training when adequate. This enables it to be applicable to different scales of data training, not limited to large-scale ones.

2.3 Multi-grained scanning

DNN are powerful in handling feature relationships, for example, Convolutional Neural Networks (CNN) are effective on image data processing, where spatial relationships among the raw pixels are critical. Recurrent Neural Networks (RNN) are effective on sequence data processing, where sequential relationships are critical. Drawn from these characteristics of Neural Networks, we enhance cascade forest with a procedure of multi-grained scanning.

As Fig. 5 illustrates, sliding windows are used to scan the raw features. Suppose that there are 400 raw features, and a window size of 100 features is used. For sequence data, a 100-dimensional feature vector will be generated by sliding the window for one feature; in total 301 feature vectors are produced. If the raw features are with spacial relationships, such as a 20 × 20 panel of 400 image pixels, then a 10 × 10 window will produce 121 feature vectors (i.e., 121 10 × 10 panels). All feature vectors extracted from positive/negative training examples are regarded as positive/negative instances, which will then be used to generate class vectors: the instances extracted from the same size of windows will be used to train a completely random tree forest and a random forest, and then the class vectors are generated and concatenated as transformed features. As Fig. 4 illustrates, suppose that there are 3 classes and a 100-dimensional window is used; then, 301

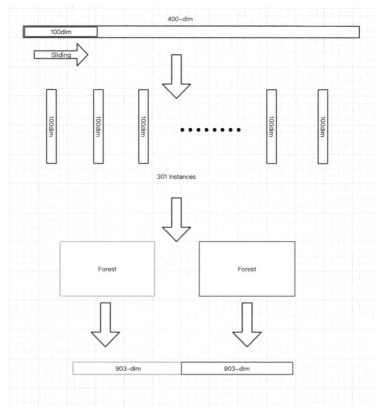

Fig. 5 Illustration of feature representation using window scanning: suppose there are three classes, raw features are 400-dim, and sliding window is 100-dim. *Source: Z.H. Zhou, J. Feng, Deep forest: towards an alternative to deep neural networks, 2017, arXiv: 1702.08835.*

three-dimensional class vectors are produced by each forest, leading to a 1806-dimensional transformed feature vector corresponding to the original 400-dimensional raw feature vector. Noting that when transformed feature vectors are too long to be accommodated, feature sampling can be performed, e.g., by subsampling the instances generated by sliding window scanning, since completely random trees do not rely on feature split selection whereas random forests are quite insensitive to inaccurate feature split selection.

Fig. 5 shows only one size of sliding window. By using multiple sizes of sliding windows, differently grained feature vectors will be generated.

3. Machine learning for detecting fraud online transaction

ML uses statistical techniques to enable computer systems to "learn" [4] (e.g., progressively improve performance on a specific task) from data, without being explicitly programmed. ML explores study and construction of algorithms that can learn from and make predictions on data—such algorithms overcome following strictly static program instructions by making data-driven predictions or decisions, through building a model from sample inputs. ML is employed in a range of computing tasks where designing and programming explicit algorithms with good performance is difficult or infeasible. Its application includes email filtering, detection of network intruders, and computer vision [4].

ML is closely related to computational statistics, which also focuses on prediction-making through the use of computers. It has strong ties to mathematical optimization, which delivers methods, theory and application domains to the field. ML include data mining, where the latter subfield focuses more on exploratory data analysis and is known as unsupervised learning [19].

ML techniques employ a prediction model based on a set of instances. The model has a parametric function, predicting the likelihood of being fraud of a transaction by a set of features describing the transaction given. Using ML in fraud detection has big potential, and has advantages: for example, it allows discovery of patterns in high dimensional data streams, i.e., transactions arrive as a continuous stream and each transaction is defined by many variables. ML techniques can be used to detect and model existing fraudulent techniques as well as identify new techniques that hint unusual behavior of the cardholders. Predictive models based on ML are also able to automatically integrate investigators' feedbacks to improve the accuracy of the detection, while in the case of expert system, including investigators feedbacks requires rules revision that can be time consuming.

Both supervised [20] and unsupervised [21] ML algorithms have been proposed for transaction fraud detection. Based on supervised learning, a model is trained by training set in order to learn the fraud patterns. Thus the supervised model could discover whether current transaction is genuine or not. On the contrary, what the unsupervised learning deals with is the unlabeled transactions. These unsupervised methods consist in outlier detection or anomaly detection techniques that associate fraudulent behaviors to any transaction that does not conform to the majority, without knowledge on transactions class. In this chapter, we focus on supervised ML algorithms.

3.1 Definition of supervised learning

Supervised Learning [20] is an important form of ML. It is named as *supervised*, because the learning process is done under the seen label of observation variables; in contrast, in Unsupervised Learning, the response variables are not available. In Supervised Learning, datasets are trained with the training sets to build ML, and then will be used to label new observations from the testing set. As for the training set, the input variables are the features which will influence the accuracy of predicted variable. It contains both quantitative and qualitative variables; the output variable is the label class that Supervised Learning will label the new observations. According to different types of output variables, Supervised Learning tasks can be divided into two kinds: classification task and regression task. The output variables of classification task are categorical variables, and those of regression task are continuous variables. For example, current hot image classification is a classification task, and prediction of stock price is a regression task.

The procedure of Supervised Learning can be described as the follows: we use $x^{(i)}$ to denote the input variables, and $y^{(i)}$ to denote the output variable. A pair $(x^{(i)}, y^{(i)})$ is a training example, and the training set that we will use to learn is $\{(x^{(i)}, y^{(i)}), i = 1, 2, ..., m\}$. (i) in the notation is an index into the training set. We use X denote the space of input values, and Y as the space of output values. The goal is, given a training set, to learn a function $h: X \rightarrow Y$ so that $h(x)$ is a good predictor for the corresponding value of y. h is a hypothesis.

3.2 Classification task

Classification task is conducted through independent variables, including continuous or discrete variables. As already mentioned in the previous sector, allowing the dependent variable $y \in Y$ takes a small set of K possible classes $Y = \{c_1, c_2, ..., c_k\}$ and the input/output value of sample (x, y) x is extracted from the joint distribution $F_{x, y}$. The classifier K is the assumption $h(x, \theta)$ to return $\hat{y} = \hat{c} \in Y$.

There are three methods to solve the classification problems:

The posterior probability $P(y = c_k | x = c_k)$ is estimated, and a class $\hat{y}$ is assigned to an input value x using the decision theory.

The class conditional probability $P(x=x|y=c_k)$ and the prior probability $P(y=c_k)$ are estimated separately, and the posterior probability is calculated using Bayes' theorem:

$$P(y = c_k|x = c_k) = \frac{P(x = x|y = c_k)P(y = c_k)}{\sum_{k=1}^{K} P(x = x|y = c_k)P(y = c_k)} \tag{10}$$

Each new x is then assigned to one of the classes by using decision theory.

Finding a function $g(x)$, discriminant function [22] is called, it will direct each input x mapped to class label $\hat{y}$.

The method of directly simulating the posterior probability is called discriminative models, and the generation method simulates the class conditional and the prior probability of each class, and then uses Bayes' theorem to calculate the posterior probability.

In the case of $K=2$ binary classification, popular loss function is zero–one loss $L_{0/1}$, which specifies loss equal to one in the case of wrong predictions, and zero otherwise:

$$L_{\frac{0}{1}} : Y \times Y \to \{0,1\} \ y, \hat{y} \to \begin{cases} 0 \ if \ y \neq \hat{y} \\ 1 \ if \ y = \hat{y} \end{cases} \tag{11}$$

However, in many applications, the cost of misclassification is dependent on the class. For example, in cancer treatment, the costs of incorrectly predicting a patient (a False Negative) are much higher than making wrong predictions when the patient is healthy (a False Positive). Assume $Y=\{1,0\}$, where 1 represents positive instances and 0 represents negative instances. Let $l_{i,j}$ be the true rank j and $p=P(y=1|x=x)$ determine my loss i (cost). We can define the risk of the predicted instance as positive or negative, as shown:

$$r^+ = (1 - p)l_{1,0} + pl_{1,1} \tag{12}$$
$$r^- = (1 - p)l_{1,0} + pl_{0,1} \tag{13}$$

Based on decision theory [12], the standard decision process defines the optimal class of the sample as the minimization of risk (the expected value of the loss function). The Bayes' decision rule used to minimize risk can be stated as follows: assign the positive class to the sample with $r^+ \leq r^-$, and the negative class otherwise.

$$\hat{y} = \begin{cases} 1 \ if \ r^+ \leq r^- \\ 0 \ if \ r^+ > r^- \end{cases} \tag{14}$$

As shown below, positive $p>\tau$ attribute and threshold τ when corresponding to the predicted sample:

$$\tau = \frac{l_{1,0} - l_{0,0}}{l_{1,0} - l_{0,0} + l_{0,1} - l_{1,1}} \tag{15}$$

Normally, the correct forecast cost is zero, thus $l_{0,\,0}=0$ and $l_{1,\,1}=0$. We obtain $L_{0/1}$ and $\tau=0.5$.

3.3 Supervised learning for fraud detection

There are four major methods in Supervised Learning algorithms: supervised profiling, classification, cost-sensitive method and network methods.

3.3.1 Supervised profiling method

It is possible to study the variable distribution of legal transactions and fraudulent transactions given available labeled transactions. Based on this premise, we could establish different profiles for different label classes and compare their similarities. For example, it is likely to use Weight of Evidence (WOE) as similarity measure between two profiles [23]. As mentioned earlier, expert rule methods are easy to understand and execute, thus we could define a set of rules for each profile. If the new issue matches the set of rules, we then consider the new issue having the same profile. However, the fraudulent techniques are fast evolving. Therefore, it is necessary to update fraudulent profiles. A weighted ensemble approach can be used to include new rules while maintaining old rules [24]. Only in this way, fraud techniques and users' behaviors can be reflected properly.

3.3.2 Classification method

ML can conduct a large number of classification tasks with good performance. In 2014, Khormuji et al. [6] presented a cascade Artificial Neural Network for detecting credit card fraud. The system aimed at attaining a high recognition rate and a high reliability. It produced good results on the database from a large Brazilian bank. In 2015, Liu et al. [7] applied four statistic methodologies, including parametric and non-parametric models to construct detection models, concluded that random forest could improve the detection efficiency significantly and have an important practical application. However, these detection methods have a very low False Positive (FP) and False Negative (FN) along with fair accuracy on fraud transaction detection task. In 2015, Ding [8] proposed a credit card transaction fraud detection model based on Deep Belief Networks (DBN), and trained a

DBN model with five layers. Kang et al. [9] applied CNN for credit card fraud detection, established a fraud transaction detection framework based on Lenet-5 structure, and achieved a good detection result. In 2017, Wang et al. [10] adopted RNN on Jingdong's real electronic transaction data, and established a detection framework and achieved over three times improvement than existing fraud detection approaches. In 2018, Zhang et al. [11] used CNN to construct the model of network transaction fraud detection with the original characteristics of transaction as input, and verifies the validity of the model in real transaction data. Despite of these progresses, deep learning requires plenty of training parameters, complicated model structure, and it is easy to cause exploding gradient or vanishing gradient. Also it may require long training time and has higher demand on the dimensions.

3.3.3 Cost-sensitive method
Cost-sensitive method could be taken into account the cost of FP and FN. Traditional cost-sensitive learning such as AdaCost [25] with assumption of a fixed costs and dependent class, is not suitable for fraud detection. Because the costs are proportional to the transaction amount, which means that the cost of FN is dynamic. Besides, among the methods, Sahin et al. [15] proposed a new cost-sensitive decision tree algorithm, which outperformed the existing well-known methods.

3.3.4 Network methods
The detection of links between data can also lead to fraud discovering. For example, a fraudulent account may have contact with some other accounts, which is like a fraud network. At the same time, we could analyze the pattern of related transactions to find the fraud activities. Especially it is possible to discover fraud groups using link analysis and graph mining methods [26]. A framework for credit card fraud detection, which is called APATE [27], shows features including network information are able to improve significantly the performance of a standard supervised algorithm.

3.4 Unsupervised learning for fraud detection
Under condition that there is no class label in transactions set, that is, we do not know whether each transaction in the set is fraudulent or not, unsupervised learning is a better process approach. Unsupervised learning can make selection independent of supervised selection. And it is possible to reveal some unseen fraudulent techniques. However, there are two core

problems which block the application of unsupervised learning in fraud detection. The one is mislabeled transactions, which will lead to invalid model. The other is the unbalanced problem. Because unsupervised learning does not know the class label, the unbalanced problem will seriously affect its effect. That is why unsupervised fraud detection has received less attention in the literature.

In unsupervised learning, outliers detection methods [28,29] are often used. Peer Group Analysis [30] put customers into different profiles and identifies frauds when transactions showing trace that are inconsistent with the customer profile. It is possible to simulate clients' behaviors by means of self-organizing maps [31].

When it is possible to define a region or distribution of the data representing the normal behavior, then all observations falling outside of those can be flagged as outliers. However, it is hard to define a normal region and the boundary between normal and outlying behavior. Besides, the normal behavior will change as time passes. Last but not least, some frauds will appear similar as the normal, while normal data also contains noise similar to frauds. These issues above are needed to be considered seriously when using unsupervised learning.

3.5 Feature engineering for fraud detection

The features in data are important to the predictive models and will influence the results. The quality and quantity of the features greatly influence the performance of the model.

Feature engineering is the process of using domain knowledge of the data to create new input features in Machine Learning algorithm. Feature engineering is fundamental to Machine Learning. It is a difficult and cost-consuming task.

Feature engineering usually includes the following steps.

➤ *Feature Extraction*: Making preliminary screening and extraction to deal with the problem of missing values and redundancy of data. Options for processing of missing values, include mean interpolation, majority interpolation, etc. Options for processing of redundancy of data include data mining, etc.

➤ *Feature Construction*: For datasets with low effective feature dimension, it is necessary to construct new features. Therefore, it is necessary to cross-assemble different features, so that the characteristics can be interconnected to each other, thus expressing the non-linearity of a single feature.

➤ *Feature Scaling*: Feature scaling is a method used to standardize the range of independent variables or features of data. In data processing, it is also known as data normalization and is generally performed during the data pre-processing step.

➤ *Feature Dimension Reduction*: Dimension reduction is the process of reducing the number of random variables under consideration by obtaining a set of principal variables. It can be divided into feature selection and feature extraction.

4. Pressing issues in detecting fraud online transaction

The fraud detection model based on ML algorithms is challenging for the following reasons:

1. Fraudulent transactions account for a small portion of all the daily transactions

   This is also known as the unbalanced problem [13]. The distributions of genuine and fraud transactions are not only unbalanced, but also overlapping. Most ML algorithms are not designed to cope with both unbalanced and overlapped class distributions.

2. Frauds distribution evolves quickly over time

   The change of fraudulent activities and costumer behavior will lead to the evolution of online transactions distribution. This situation is typically referred to as concept drift and is of extreme relevance for fraud detection which has to be constantly updated either by exploiting the most recent supervised samples or by forgetting outdated information that might be no more useful whereas not misleading.

3. The true transaction label will be only available after several days, because investigators could not timely check all the transactions

   Investigators tend to choose limited high risk transactions to check, which will lead to mislabeling of the transactions. The results investigators checked will be used as feedback labels to retrain or update the fraud detection model, so resolving this challenge is very important.

As for the mentioned challenges, there are several methods to deal with.

4.1 Sampling methods

Typically, sampling methods are used to rebalance the datasets, because standard classifiers have better performances when trained on a balanced training set. Sampling techniques do not take into consideration any class information in removing or adding observations, yet they are easy to implement and to understand.

The way to deal with data imbalance usually includes the methods of under-sampling, oversampling, Synthetic Minority Oversampling Technique (SMOTE) and so on. Each method has its own advantages and disadvantages and scope of application.

Under-sampling [32] is to narrow most classes by randomly removing observations. In an unbalanced problem, it is realistic to assume that many of the observations of most classes are superfluous, so by randomly deleting some of them, the distribution of results should not be changed too much. However, the risk of removing the observed from the data set persists because the removal is carried out in an unsupervised manner.

Oversampling [32] is to adjust the size of a sample to reduce the degree of imbalance in the sample class. By copying a few classes until two classes have equal frequencies. Oversampling increases the risk of overfitting by biasing the model to a few classes.

SMOTE [33] oversamples the minority class by generating synthetic examples in the neighborhood of the observed ones. The idea is to form new minority examples by interpolating between samples of the same class. SMOTE can improve the performances of a base classifier in many applications, but it has also some drawbacks: Synthetic observations are generated without considering neighboring examples, leading to an increase of overlap between the two classes.

4.2 Cost-based methods

In unbalanced classification tasks, it is usually more important to correctly predict positive (minority) instances than negative instances. This is often achieved by associating different costs to erroneous predictions of each class. Cost-based methods operating at the algorithm level are able to consider misclassification costs in the learning phase without the need of sampling the two classes.

In the family of decision tree classifiers, cost-based splitting criteria are used to minimize costs [34], or cost information determines whether a sub-tree should be pruned [35]. In general, pruning allows improving the generalization of a tree classifier since it removes leaves with few samples on which we obtain too low probability to evaluate the instance.

5. Case study: Deep forest in detecting online transaction fraud

Traditional techniques for detecting fraudulent transactions, such as rule-based expert system, are easy to be understood, but cannot detect concealed fraud transactions and have the risk of rule invalidity. Online fraudulent transactions are always difficult to analyze, as they are designed to avoid detection. To meet the challenge, we propose a gcForest-based approach for online transaction fraud detection in this session.

We propose a differentiation feature generation method based on transaction time, where a customer transaction behavior is observed to derive some features using a transaction aggregation strategy. The derivation of aggregation features is to group transaction records over a given period of time. We also adopt the aggregation feature strategy of sliding time window, select the transaction records made in the past time and calculate the number of transactions and the amount of transactions on the corresponding time window. The information of a single transaction is not sufficient to detect a fraudulent transaction, as it does not contain some important information, such as the consumer behavior. Therefore, we choose to adopt a transaction time-based differentiation feature generation method. Individual Credibility Degree (ICD) and Group Anomaly Degree (GAD) based on transaction time are derived to distinguish legal transactions and fraudulent transactions further.

Meanwhile, to deal with the extreme imbalance of fraudulent transactions, we apply deep-forest algorithm to detect fraudulent transactions. While raw deep-forest model could ignore the outlier transaction samples, we enhance the raw deep forest with detection mechanism for outliers, paying more attention on outliers to promoting the precision of fraud detection model. Also, resampling methods significantly improve the performance of gcForest-based fraud detection.

Feedbacks from investigators provide recent supervised samples that are highly informative. Here we proposed a gcForest-based Fraud Transaction Detection Framework, and in our scheme, the below two methods are combined to automatically detect the fraudulent transactions and send feedback to investigators.

5.1 GcForest-based fraud transaction detection framework

GcForest-based fraud transaction detection framework consists of model training module and detection module, which is shown in Fig. 6.

In the model training section, we firstly pre-process the training data to improve the quality of the input data. Then we derive the differentiation features from available attributes in the transaction data. On the one hand, a time window-based feature aggregation strategy is adopted to excavate the patterns of the customer's normal transactions. On the other hand, ICD and

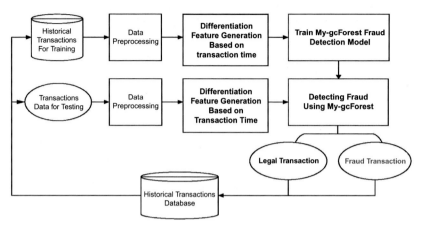

Fig. 6 gcForest-based transaction fraud detection framework. *Source: Author.*

GAD based on transaction time are generated to find the difference between legal transactions and fraudulent transactions. Finally, we input the training data into gcForest–based model with outlier samples detection to train a fraud detection classifier.

In the detection module, we employ the same procedures as model training module in the first two steps for the testing data. Afterwards, we input the treated testing data into the fraud detection model training by model training part, which determines whether the testing data is a fraudulent transaction or not.

5.2 Transaction time-based differentiation feature generation method

To better distinguish legal transactions and fraudulent transactions, we propose a differentiation feature generation method based on transaction time. A more appropriate solution taken into account customer transaction behaviors is to derive some features using a transaction aggregation strategy. To derive aggregation features, we group transaction records over a given period of time. We adopt the aggregation feature strategy of sliding time window, selecting the transaction records made in the past time t_p and calculating the number of transactions and the amount of transactions on the corresponding time window t_p.

However, it is not enough to convey a customer's all the information using aggregated features. Transaction time is a very important transaction feature for online transactions, but the discretization of attribute values is too large to be used directly, which has a great impact on the classification model. When analyzing the transaction time, it is easy to make the mistake of using the arithmetic mean. Indeed, the arithmetic mean is not a correct way to average time, because it does not take into account the periodic behavior of the time feature [36].

The von Mises distribution [37], also called circular normal distribution, can reflect the distribution characteristics of periodic variables. Considering the periodic behavior of the time feature, we could map the transaction time characteristics to a 24-h circular timeline. The von Mises distribution for the transaction time set $D = \{t_1, t_2, \ldots, t_n\}$ is defined as follows:

$$D \sim vonmises\left(\mu, \frac{1}{\sigma}\right) \tag{16}$$

where μ is the periodic mean and σ is the periodic standard deviation.

On this basis, we propose ICD and GAD based on transaction time for distinguishing legal transactions and fraudulent transaction. ICD is individual credibility degree of the customer's current transaction time considering distribution of the normal transaction time of a customer in the past period of time. GAD is group anomaly degree of the customer's current transaction time considering distribution of all the fraudulent transactions in the past period of time.

Let S be a set of transactions with size $= M$, $M = |S|$. For current transaction i of a customer number x_i^{id}, consider a set of transaction time within the past time t_p.

$$Time = \left\{ x_l^{time} \middle| \left(x_l^{id} = x_i^{id}\right) \bigwedge \left(hours\left(x_i^{time}, x_l^{time}\right) < t_p\right) \right\}_{l=1}^{M} \tag{17}$$

where x_l^{time} represents the transaction time of transaction l, x_i^{id} represents the customer number of transaction l, x_i^{time} represents the current transaction time of the customer number x_i^{id} and $hours(x_i^{time}, x_l^{time})$ is a function used to calculate the time difference between x_i^{time} and x_l^{time}.

Then cluster the transaction time set, and the cluster number is k. Using k-means clustering algorithm, we get k transaction time clusters $time_i$ $(i = 1, 2, \ldots, k)$. At the same time, calculate each cluster's the number of transaction time as $number_i (i = 1, 2, .., k)$, and the weight is given to each cluster $time_i$:

$$weight_i = \frac{number_i}{N} \quad i = 1, 2, \ldots, k \tag{18}$$

where $N = |Time|$ represents the size of transaction time.

Next, we calculate the transaction time probability for k clusters:

$$x_i^{time} \sim \text{von Mises}\left(\mu(time_i), \frac{1}{\sigma(time_i)}\right) i = 1, 2, ..., k \tag{19}$$

where $\mu(time_i)$ represents the periodic mean of the transaction time each cluster $time_i$, and $\sigma(time_i)$ represents the periodic standard deviation of the transaction time each cluster $time_i$.

Meanwhile, the confidence degree α is defined. The confidence interval for the transaction time for each cluster $time_i$ is calculated as follows:

$$\left[x_i^{time} - \frac{\sigma(time_i)}{\sqrt{number_i}} z_{\frac{\alpha}{2}}, x_i^{time} + \frac{\sigma(time_i)}{\sqrt{number_i}} z_{\frac{\alpha}{2}}\right] i = 1, 2, ..., k \tag{20}$$

We observe whether the actual current transaction time is within the confidence interval of each transaction time cluster $time_i$. If the actual transaction time is within the confidence interval, let $p_i = 1$ otherwise $p_i = 0$. We calculate ICD as follow:

$$ICD = \sum_i^k p_i * weight_i \; i = 1, 2, ..., k \tag{21}$$

At the same time, fraudsters tend to avoid the normal transaction hours of customers and the fraudulent group behaviors, so we propose the concept of GAD. Extract the set of all fraudulent transaction time, and the previous steps are taken for the set to calculate GAD:

$$GAD = \sum_j^K p_j * value_j \; j = 1, 2, ..., K \tag{22}$$

where K is the number of clusters in the aggregate of fraudulent transactions, $value_j$ is the weight given to each cluster j after clustering.

Finally, we get the confidence degree feature of the current transaction time:

$$confidence = ICD - GAD \tag{23}$$

5.3 GcForest-based model with outlier samples detection

In this section, we propose a gcForest-based approach to detect online transaction fraud with outliers detection mechanism. The raw gcForest generates a deep-forest ensemble, with a cascade structure which enables gcForest to do representation learning. Besides, random forest used in each layer of the cascade structure has a good processing mechanism for imbalanced samples with the bagging mechanism [38]. However, due to the extreme imbalance of fraudulent transactions, the outlier samples that need to be focused on are easy to be neglected. It is likely to fail to detect some transaction samples, reducing the effectiveness of the model on detecting fraud transactions.

Here, we propose an outlier samples detection mechanism in gcForest-based model. As shown in Fig. 7, all transaction records in the initial training set are given the same weights, thus the samples have the same probability of being sampled. Next, we train the classifier. For the correctly classified samples, we change their weights and make them possible to extract with a lower probability in the next procedure. For the misclassified samples, in other words, the outlier samples, changing their weights makes the latter model pay more attention to them. This promotes the precision of fraud detection model. This mechanism is the core of boosting, an ensemble learning method [39].

We adopt the cascade architecture of raw gcForest, replacing the completely random forest with XGBoost [40] (eXtreme Gradient Boosting). The XGBoost model has advantages of dealing with unbalanced samples, calculating fast and treating table data with less variables better. At the same time, it can reduce overfitting and calculation. XGBoost model has been proved to be practical for many classification and regression tasks [41].

As illustrated in Fig. 8, New-gcForest employs a cascade structure, where each layer of the cascade forest receives feature information processed by its preceding layer and outputs its processing result to the next layer. Each layer of the cascade forest sets the same number of random forest and xgboost.

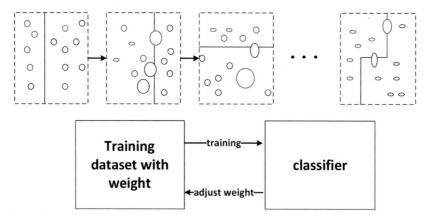

Fig. 7 The outlier samples detection mechanism. *Source: Author.*

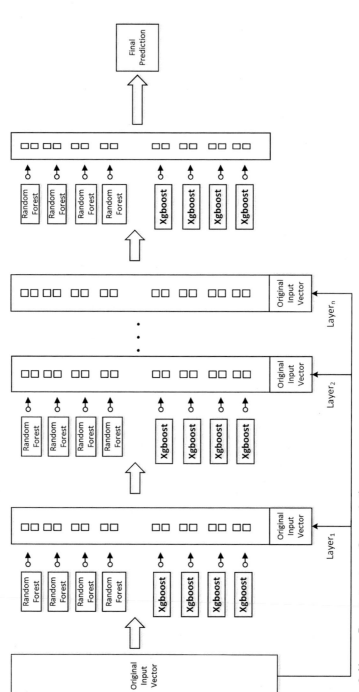

Fig. 8 New-gcForest structure. *Source: Author.*

After extending a new layer, the performance of the entire cascaded structure is evaluated on the validation set. If there is no significant increase in performance, the entire training process will end. Therefore, the number of cascaded layers can be adaptively determined, such that the model complexity can be automatically set, enabling gcForest to perform excellently even on small-scale data [14]. Because of low dimensional data we use in the experiment, results show that the multi-grained scanning of raw gcForest makes no contribution to model performance, so we abandon it. To reduce the risk of overfitting, we use k-fold cross validation for generating class vectors produced by random forest and XGBoost.

5.4 Experiment and evaluation

Our experiment data comes from a Chinese bank's real online transaction data. It includes 3-month B2C (Business-to-Customer) transaction records. There are original eight available transaction attributes, and more than 70,000 transactions labeled as fraudulent transactions in experimental data. In our experiment, we use transaction data of the first 2 months as a training set to train the gcForest-based fraud detection model. The last month's transaction data is used as the testing set to evaluate the performance of the detection model.

We pre-process the data from real transaction records, and adopt differentiation feature generation method for the 3 months of transaction data. 17-dimensional features are derived from the method, thus the input feature number of the model is 25. Then 2 million transaction records from the transaction data of the previous 2 months are sampled as the training set, which is used to train New-gcForest model. Six testing sets (200,000, 400,000, 600,000, 800,000, 1,000,000, 1,200,000) are selected from the transaction data of the third month, which are used to evaluate the effectiveness of detection model. Besides, we employ precision and recall to evaluate the performance of fraud detection model.

We compare our *New-gcForest* with the existing methods, *random forest* model and *raw gcForest* model. The existing gcForest based on random forest applies the sampling methods with replacement, while the samples of minority class, which should be paid more attention to, are easy to be neglected owing to the extreme imbalance of fraudulent transactions. In contrast to the existing one, New-gcForest adopts the outliers mechanism, which focuses on the samples of minority class by giving weights. The performance of New-gcForest is better than other two methods according to our experiment.

As shown in Fig. 9, with the six testing sets adopt the same data pre-processing and use the same features as model input, New-gcForest model

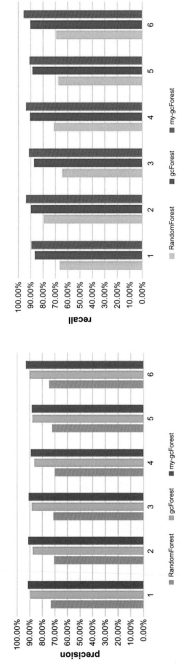

Fig. 9 Evaluation of fraud detection model on six testing sets. *Source: Author.*

achieves the best detection results, improving precision rate by 15% and recall rate over 20%, in comparison with random forest detection model. In contrast to raw gcForest, New-gcForest has an average increase of 3–5% in precision and recall rate.

To evaluate the performance of multi-grained scanning in raw gcForest, we compare two models on three testing sets, including raw gcForest model with multi-grained scanning and raw gcForest model without multi-grained scanning. The three testing sets adopt the same data pre-processing and use the same features as model input. Experiments show that the multi-grain scanning module does not significantly improve the detection effect. Therefore, we abandon the multi-grained scanning part of raw gcForest. The contrast experimental results are shown in Fig. 10.

Fig. 11 shows that the confidence degree feature derived from the differentiation feature generation method helps to promote the performance of fraud detection model. We adopt raw gcForest model and New-gcForest model to evaluate the importance of confidence degree feature. Experiments on four testing sets demonstrate that the confidence degree feature enhances the effect of fraud detection model to some extent, which increases the precision and recall of the model by an average of about 5%.

5.5 Future research

The main contributions to this chapter can be summarized as follows:
➢ We propose a gcForest-based approach to detect online fraudulent transactions.
➢ We employ ICD and GCD based on transaction time, to capture the internal relationship between transaction attributes and fraud patterns.
➢ We improve cascade forest of the raw gcForest and create a detection mechanism for outliers to further enhance the effect of fraud detection.

A standard approach to deal with unbalanced classification tasks is to rebalance the dataset before training a learning algorithm. The outliers detection mechanism of our New-gcForest can alleviate the problems caused by unbalanced samples to some extent, but it cannot totally solve this problem. There is a lot of researchers work on methods to deal with unbalanced data. However, a definitive solution has not yet been found. We need to explore more sophisticated sample imbalance solutions further.

Besides, there can be further exploration on higher-dimensional data to distinguish legal transactions and fraudulent transactions. We could discover more interesting issues from high dimensional data, because the more

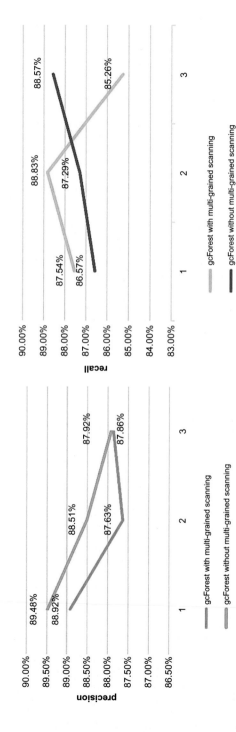

Fig. 10 Evaluation of multi-grained scanning module on three testing sets. *Source: Author.*

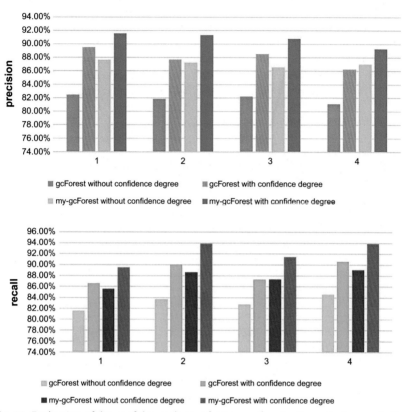

Fig. 11 Evaluation of the confidence degree feature on four testing sets. *Source: Author.*

information you get, the more you could analyze. Therefore, it is possible to use representation learning to acquire more useful information from transaction data, so deep learning could be the solution. We could apply deep learning algorithms including CNN, RNN and Auto encoder to gain more feature information beside the transaction.

6. Summary of key lessons learned

Through analysis on deep forest for detecting online transaction fraud, we learned the following lessons:

➢ Under-sampling technology is not guaranteed to improve the performance of model. There are several factors which influence the effectiveness of under-sampling and most of these factors could not be controlled.

➢ The rate of sampling transaction data depends on the specific datasets. Also, the imbalance ratio will influence the performance of the model and the effectiveness of the sampling methods.

➢ The non-stationary distributions of data stream exist in online transactions. It is significant to retain historical transactions as well as forget outdated samples for the model when using samples for training.

7. Conclusion

In this chapter, we demonstrate the fundamentals of ML's application in fraud detection and analyze the key design considerations for online transaction fraud detection. Through the case study of one ML algorithm, Deep forest, we examine the ways to meet the challenges in detecting online transaction fraud. The problems of big computing load and unbalanced distribution still remain in improving the fraud detection system. We introduce a transaction time-based differentiation feature ICD and GAD generation method into our scheme. Furthermore, to deal with the extreme imbalance of online transactions, we apply Deep-forest algorithm to detect fraudulent transactions. Furthermore, we enhance the raw Deep forest with detection mechanism for outliers, paying more attention on outliers to promote the precision of fraud detection model. Finally, we conduct test using one bank's transaction data. Compared with random Forest-detection model, our method improves precision rate by 15% and recall rate by 20%.

Acknowledgments

This work was supported by the Natural Science Foundation of Shanghai under Grant 19ZR1401900, Shanghai Science and Technology Innovation Action Plan Project under Grant 19511101802, and National Natural Science Foundation of China under Grant 61472004 and Grant 61602109.

Key terminology and definitions

Expert driven system Expert driven system is a fraud detection system based on domain knowledge from fraud investigators to define rules to predict the probability of a new transaction to be fraudulent.

Fraud Detection System Fraud Detection System is a system controlling whether the transactions are legal or should be reported as a fraud. The system usually contains five elements: (1) Terminal, (2) Transaction Blocking Rules, (3) Scoring Rules, (4) Data Driven Model, (5) Investigators. The first four elements are fully automatized, while the last one requires human intervention and it is the only non-automatic and offline part of Fraud Detection System.

von Mises distribution In probability theory and directional statistics, the von Mises distribution (also known as the circular normal distribution or Tikhonov distribution) is a continuous probability distribution on the circle. It is a close approximation to the wrapped normal distribution, which is the circular analogue of the normal distribution. A freely diffusing angle θ on a circle is a wrapped normally distributed random variable with an unwrapped variance that grows linearly in time. On the other hand, the von Mises distribution is the stationary distribution of a drift and diffusion process on the circle in a harmonic potential, i.e. with a preferred orientation. The von Mises distribution is the maximum entropy distribution for circular data when the real and imaginary parts of the first circular moment are specified. The von Mises distribution is a special case of the von Mises–Fisher distribution on the N-dimensional sphere.

The Non-stationary Distributions/Concept Drift In predictive analytics and machine learning, the concept drift means that the statistical properties of the target variable, which the model is trying to predict, change over time in unforeseen ways. This causes problems because the predictions become less accurate as time passes. The term concept refers to the quantity to be predicted. More generally, it can also refer to other phenomena of interest besides the target concept, such as an input, but in the context of concept drift, the term commonly refers to the target variable.

Cross-validation Cross Validation, sometimes called Rotation Estimation, is a practical way to statistically cut data samples into smaller subsets, proposed by Seymour Geisser. In a given modeling sample, take most of the samples to build the model, leaving a small part of the sample to be forecasted with the model just created, and the prediction error of this small part of the sample, and record their square sum. This process continues until all samples are forecasted once and only once. The squared error of the prediction error of each sample is called PRESS (predicted Error Sum of Squares).

References

[1] R. Anderson, The Credit Scoring Toolkit: Theory and Practice for Retail Credit Risk Management and Decision Automation, Oxford University Press, New York, 2007.

[2] A. Dal Pozzolo, Adaptive machine learning for credit card fraud detection, PhD thesis, Universite Libre de Bruxelles, 2015.

[3] China Leiwang Platform, Cyber Fraud Trend Research Report (in Chinese), 2017, Available from http://zt.360.cn/1101061855.php?dtid=1101062366&did=491006041. 2018.

[4] A.L. Samuel, Some studies in machine learning using the game of checkers, IBM J. Res. Dev. 3 (1959) 210–229.

[5] T.P. Bhatla, V. Prabhu, A. Dua, Understanding credit card frauds, TCS 1 (2003) 1–14.

[6] M.K. Khormuji, M. Bazrafkan, M. Sharifian, S.J. Mirabedini, A. Harounabadi, Credit card fraud detection with a cascade artificial neural network and imperialist competitive algorithm, Int. J. Comput. Appl. 96 (2014) 1–9.

[7] C. Liu, Y. Chan, S.H. Alam Kazmi, H. Fu, Financial fraud detection model based on random forest, Int. J. Econ. Financ. 7 (7) (2015) 178–188.

[8] W.X. Ding, Research on credit card transaction fraud detection based on deep learning technology, Master thesis, Shanghai Jiaotong University, 2015.

[9] K. Fu, D. Cheng, Y. Tu, L. Zhang, Credit card fraud detection using convolutional neural networks, in: International Conference on Neural Information Processing, 2016, pp. 483–490.

[10] S. Wang, C. Liu, X. Gao, H. Qu, W. Xu, Session-based fraud detection in online e-commerce transactions using recurrent neural networks, in: Joint European Conference on Machine Learning and Knowledge Discovery in Databases, 2017, pp. 241–252.

[11] Z.H. Zhang, X.X. Zhou, X.B. Zhang, L.Z. Wang, P.W. Wang, A model based on convolutional neural network for online transaction fraud detection, Secur. Commun. Netw. 2 (2018) 1–9.

[12] R.J. Bolton, D.J. Hand, Unsupervised profiling methods for fraud detection, in: Conference on Credit Scoring and Credit Control, 2001, pp. 235–255.

[13] H. He, E.A. Garcia, Learning from imbalanced data, IEEE Trans. Knowl. Data Eng. 21 (2009) 1263–1284.

[14] Z.H. Zhou, J. Feng, Deep forest: towards an alternative to deep neural networks, Proc. Twenty-Sixth Int. Joint Conf. Artif. Intell. (2017) 3553–3559, https://doi.org/10.24963/ijcai.2017/497.

[15] Y. Sahin, S. Bulkan, E. Duman, A cost-sensitive decision tree approach for fraud detection, Expert Syst. Appl. 40 (2013) 5916–5923.

[16] A. Criminisi, J. Shotton, E. Konukoglu, Decision forests: a unified framework for classification, regression, density estimation, manifold learning and semi-supervised learning, in: Foundations and Trends in Computer Graphics and Vision, vol. 7, NOW Publishers, 2012, pp. 181–227.

[17] L. Breiman, Random forests, Mach. Learn. 45 (1) (2001) 5–32, https://doi.org/10.1023/A:1010933404324.

[18] F.T. Liu, K.M. Ting, Y. Yu, Z. Zhou, Spectrum of variable-random trees, J. Artif. Intell. Res. 32 (2008) 355–384.

[19] G.E. Hinton, T.J. Sejnowski, Unsupervised Learning: Foundations of Neural Computation, MIT Press, 1999.

[20] S. Bhattacharyya, S. Jha, K. Tharakunnel, J.C. Westland, Data mining for credit card fraud: a comparative study, Decis. Support. Syst. 50 (2011) 602–613.

[21] D.K. Tasoulis, N.M. Adams, D.J. Hand, Unsupervised clustering in streaming data, in: ICDM Workshops, 2006, pp. 638–642.

[22] C.M. Bishop, Pattern Recognition and Machine Learning, Springer Press, 2016.

[23] N. Siddiqi, Credit Risk Scorecards: Developing and Implementing Intelligent Credit Scoring, SAS Publishing, 2005.

[24] H.X. Wang, W. Fan, P.S. Yu, J. Han, Mining concept-drifting data streams using ensemble classifiers, in: Proceedings of the Ninth ACM SIGKDD International Conference on Knowledge Discovery and Data Mining, 2003, pp. 226–235.

[25] W. Fan, S.J. Stolfo, J.X. Zhang, P.K. Chan, Adacost: misclassification cost-sensitive boosting, in: ICML, 1999, pp. 97–105.

[26] A. Sudjianto, S. Nair, M. Yuan, A.J. Zhang, D. Kern, F. CelaDiaz, Statistical methods for fighting financial crimes, Technometrics 52 (2010) 5–19.

[27] V.V. Vlasselaer, C. Bravo, O. Caelen, T.E. Rad, L. Akoglu, M. Snoeck, et al., APATE: a novel approach for automated credit card transaction fraud detection using network-based extensions, Decis. Support. Syst. 75 (2015) 38–48.

[28] Z. Zhang, J. Cui, Agile perception method for behavior abnormity in large-scale network service system, Chin. J. Comput. 40 (2017) 503–519.

[29] Z. Zhang, L. Ge, P. Wang, X. Zhou, Behavior reconstruction models for large-scale network service systems, Peer Peer Netw. Appl. 12 (2019) 502–513.

[30] D.J. Weston, D.J. Hand, N.M. Adams, C. Whitrow, P. Juszczak, Plastic card fraud detection using peer group analysis, Adv. Data Anal. Classif. 2 (2008) 45–62.

[31] J.T.S. Quah, M. Sriganesh, Real-time credit card fraud detection using computational intelligence, Expert Syst. Appl. 35 (2008) 1721–1732.

[32] C. Drummond, R.C. Holte, C4.5, class imbalance, and cost sensitivity: why under-sampling beats over-sampling, in: Workshop on Learning From Imbalanced Datasets II, 2003.

[33] N.V. Chawla, K.W. Bowyer, L.O. Hall, W.P. Kegelmeyer, Smote: synthetic minority over-sampling technique, J. Artif. Intell. Res. 16 (2002) 321–357.

[34] C.X. Ling, Q. Yang, J.N. Wang, S.C. Zhang, Decision trees with minimal costs, in: Proceedings of the Twenty-First International Conference on Machine Learning, 2004, p. 69.

[35] J.P. Bradford, C. Kunz, R. Kohavi, C. Brunk, C.E. Brodley, Pruning decision trees with misclassification costs, in: Machine Learning: ECML-98, 1998, pp. 131–136.

[36] A.C. Bahnsen, D. Aouada, A. Stojanovic, B. Ottersten, Feature engineering strategies for credit card fraud detection, Expert Syst. Appl. 51 (2016) 134–142.

[37] N.I. Fisher, Statistical Analysis of Circular Data, Cambridge University Press, 1995.

[38] L. Breiman, Bagging predictors, Mach. Learn. 24 (1996) 123–140.

[39] R.E. Schapire, Y. Freund, P. Bartlett, W.S. Lee, Boosting the margin: a new explanation for the effectiveness of voting methods, in: Fourteenth International Conference on Machine Learning, Morgan Kaufmann, 1997, pp. 322–330.

[40] T. Chen, C. Guestrin, Xgboost: a scalable tree boosting system, in: Proceedings of the 22nd ACM Sigkdd International Conference on Knowledge Discovery and Data Mining, 2016, pp. 785–794.

[41] D. Nielsen, Tree boosting with XGBoost-why does XGBoost win every machine learning competition, MS thesis, Norwegian University of Science and Technology, 2016.

About the authors

Lizhi Wang was born in 1994. He is a MS candidate of Donghua University. His research area includes machine learning, deep learning, and cloud computing.

Dr. Zhaohui Zhang obtained Bachelor degree in Computer Science from Anhui Normal University, Wuhu, China, in 1994. He obtained his PhD degree in Computer Science from Tongji University, Shanghai, China, in 2007. From 1994 to 2015, he worked in Anhui Normal University as professor. From 2015 to present, he works as a Professor in School of Computer Science and Technology, Donghua University, Shanghai, China. His research interests include network information services, service computing and cloud computing.

Xiaobo Zhang was born in 1989. He obtained his MS degree in Donghua University, Shanghai, China. His research interests include big data distributed system, big data query optimization and cloud computing.

Xinxin Zhou was born in 1993. She is a MS candidate of Donghua University. Her research area includes machine learning, deep learning, and cloud computing.

Dr. Pengwei Wang received BS and MS degrees in Computer Science from Shandong University of Science and Technology, Qingdao, China, in 2005 and 2008. He obtained his PhD degree in Computer Science from Tongji University in 2013. He finished his postdoctoral research work at the Department of Computer Science, University of Pisa, Italy in 2015. Currently, he is an Associate Professor with the School of Computer Science and Technology, Donghua University. His research interests include service computing, cloud computing and Petri nets.

Dr. Yongjun Zheng has successfully completed a PhD, MSc and BSc in Computer Science at Nottingham Trent University, he is employed as a Senior Lecturer in Computer Science in School of Physics, Engineering and Computer Science, University of Hertfordshire now and used to work in York St John University; he also worked as a researcher at the University of Cambridge and Middlesex University before joined York St John University. His primary research interest concerns HCI including data visualization, mobile computing and intelligent systems. His has published 26 research papers for leading international conferences and journals, including IEEE Conference on Visual Analytics Science and Technology (VAST), IEEE Transactions in Computer Graphics and Visualization, one of his research work was the winner of IEEE VAST Challenge: Subject Matter Expert Award 2012. Dr Yongjun has been the co-investigator of few funding projects. He is also reviewer of a few International Journals and is the program committee member of several International Conferences. He led and co-organized workshops in International Conferences.

CHAPTER TWO

Design of cyber-physical-social systems with forensic-awareness based on deep learning

Bin Yang, Honglei Guo, and Enguo Cao
School of Design, Jiangnan University, Wuxi, Jiangsu Province, PR China

Contents

Advances in Computers, Volume 120
ISSN 0065-2458
https://doi.org/10.1016/bs.adcom.2020.09.001

Abstract

Cyber-physical-social systems (CPSSs) contain integrated cyber parts, including computing, communication, and physical parts; it uses computations and communication
embedded in and interacting with physical processes to add new capabilities to physical systems. While CPSSs opens new opportunities in various fields, it brings new challenges in security. By inventing anti-forensic tools, attackers may break the investigation
system and make the analysis of attack scenarios difficult. In this chapter, we propose a
scheme of forensics-awareness to support reliable image forensics investigations. The
forensic scheme utilizes a Multichannel Convolutional Neural Network (MCNN) to learn
hierarchical representations from the input images. Fake images can be identified using
the well-trained MCNN. Most existing works aim at detecting a certain manipulation,
which may usually lead to misled results, e.g. results with irrelevant features and/or classifiers. To overcome this limitation, we extract the periodicity property and filtering
residual feature from the image blocks. The *f-map* is then generated by linear blending
the periodic spectrum and the residual map, and is fed into the MCNN. The overall
framework is capable of detecting different types of image manipulations, including
cloning, removing, splicing and smoothing. Experimental results show that this scheme
outperforms formerly existing ones.

Abbreviations

AI	Artificial Intelligence
ANN	Artificial Neural Network
CFA	Color Filter Array
CNN	Convolutional Neural Network
CPSSs	Cyber-Physical-Social Systems
DCT	Discrete Cosine Transform
EM	Expectation-Maximization
FPR	False Positive Rate
FRF	Filtering Residual in Frequency
GAD	Group Anomaly Degree
ICD	Individual Credibility Degree
LSTM	Long Short-Term Memory
MFM	Multichannel Feature Map
MLP	Multilayer Perceptron
NFC	Near Field Communication
NIN	Network In Network
PCA	Principal Component Analysis
PRNU	Photo-Response Non-Uniformity
QF	Quality Factor
RFID	Radio Frequency Identification
RWS	Random Walker Segmentation
SIFT	Scale Invariant Feature Transform
SURF	Speeded-Up Robust Features
SVM	Support Vector Machine
TPR	True Positive Rate
WSN	Wireless Sensor Network

1. Introduction

The cyber–physical–social systems (CPSSs), where the entities in physical world become virtual entities in cyber world, enhances both physical and digital entities' capacity in sensing, processing, and self-adapting [1]. In CPSSs, security has emerged as a big challenge. Among the security protection system, image forensic is one of the most important and fast-moving field. Image forensic is a technique to identify the integrity and authenticity of image by its statistical characteristics.

With various sophisticated editing software, people can easily make change and manipulate digital content. For example, on July 9, 2018, the Financial Times, the Los Angeles Times and other major news channels published a photo (Fig. 1A), sourcing from Sepah News, the media of Iran's Revolutionary Guard. However, The Associated Press received another image, Fig. 1B, from the same source 1 day later. Fig. 1A was then found to be tampered [2]. Digital images are becoming more and more unreliable as records of events. Other modern sensor technologies, such as Radio Frequency Identification (RFID) [3], Near Field Communication (NFC) [4], and Wireless Sensor Network (WSN) [2], also make the protection of digital content more and more complicated. Tools to authenticate digital images are in need in many fields. Our study focuses on digital forensic investigation in the CPSSs, attempting to establish a scheme to effectively verify image authenticity and detect forgery.

The remainder of this book chapter is organized as follows: Section 2 demonstrates the pressing issues in digital forensic by categorizing threats of image forgery, briefing current forensic methods, in particular some detection methods of major type of forgery. Section 3 presents the research progress and research gap of forgery detection with Deep Learning. Section 4 presents a universal system of forgery detection designed by us; Section 5 presents the experiment results of our scheme; Section 6 summaries the key contributions of our research; Section 7 points out the research directions in the field; Section 8 is conclusion of this chapter.

2. Issues in discussion

This section demonstrates pressing issues in digital forensic today, by a close look at threats of image forgery, current forensic methods, and major type forgery detections.

Fig. 1 An example of copy-move forgery. *Source: G.K. Birajdar, V.H. Mankar, Digital image forgery detection using passive techniques: a survey, Digit. Investig., 10 (2013) 226–245.*

2.1 Threats of image forgery

Image forgery threats can be grouped into three types [5]:

➤ Image retouching or enhancing

This type of forgery enhances the image features. It can be used to alter color or texture of the target area or change weather conditions (see Fig. 2 as an example).

➤ Image splicing or compositing

This type of forgery splices of two or more images into a single (see Fig. 3 for example).

➤ Clone or copy-move

This type of forgery is to erase or add some objects in the image (see Fig. 4 for example).

Apart from the above forgery methods, in many cases, additional transformations are used in order to cover forgery of the image, for example, rotation, noise addition, blurring, etc.

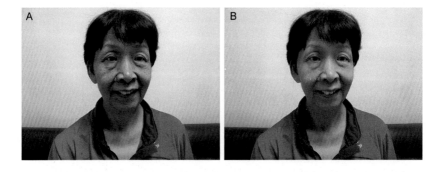

Fig. 2 An example of enhancing: wrinkles reduced in the target area. *Source: Author.*

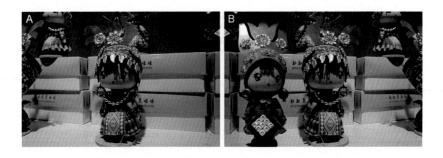

Fig. 3 An example of splicing. *Source: Author.*

Fig. 4 An example of clone. *Source: B. Yang, H. Guo, An efficient forensic method based on high-speed corner detection technique and SIFT descriptor, in: X. Sun, A. Liu, H.-C. Chao, E. Bertino (Eds.) International Conference on Cloud Computing and Security, Springer International Publishing, Nanjing, 2016, pp. 450–463.*

2.2 Image forensic methods

The forensic methods can be grouped into two categories, active forensics and passive or blind forensics [6] (see Figs. 5 and 7). Active forensic is to modify the multimedia signal prior to its distribution, in order to assist later forensic analysis. Digital watermark [7–9] is one of the examples of active forensics. Passive or blind forensic detection assesses the authenticity or

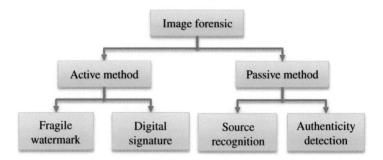

Fig. 5 The methodology of image forensic. *Source: Author.*

integrity of images without referring to signature or watermark of the original image. It assumes that although forgeries may leave no visual clue, they alter the underlying property or consistency of an image. The inconsistencies can be used to detect the forgery.

Previous forensic schemes are mainly based on the active watermark. Fragile watermark [10] and digital signature [11] had provided many methods to verify digital tampering. However, these methods are performed by embedding verification information into the original image before the image being used. A limitation of active forensics is the need for content-generating devices, e.g. cameras, sensors, microphones. Such devices are often not available, and in these cases active forensics cannot be applied. Thus, we focus on studying the passive forensic techniques in this chapter.

Passive forensic can roughly be grouped into two categories, source recognition (see Fig. 6) and authenticity detection. Modern multi-digital devices, such as camera, scanner and computer, can generate digital images. In general, source recognition relies on the underlying characteristics of the components of digital devices. These characteristics may take the form of image artifacts, distortions, and statistical properties of the underlying data. They are usually imperceptible to the human eye, but still leave clues for identification. The goal of device identification is to identify the model and/or manufacturer of the device that produced the image in question. For digital cameras, the image acquisition pipeline is usually considered [12]. Characteristics such as lens, size of the sensor, choice of Color Filter Array (CFA) are usually be used to provide forensic clues. As presented in Fig. 7, image source recognition is to identify an image's equipment in the case of unknowing the image source.

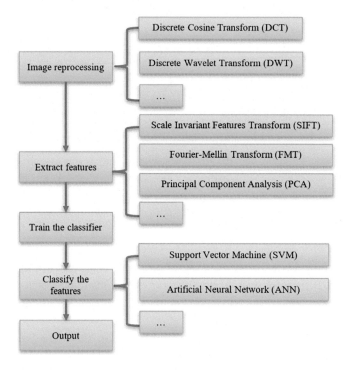

Fig. 6 The flowchart of source recognition. *Source: Author.*

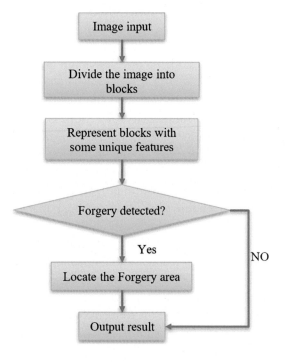

Fig. 7 Flowchart of passive image forensic detection approach. *Source: Author.*

Table 1 Highlight and weakness of different types of forensic methods.

Methods	Highlight	Weakness
Fragile watermark	Watermark need to be embedded into the original image before the image being used. Tempering operation usually destroys the integrity of the watermark, so forgery can be detected by testing the integrity	Watermark must be inserted at the time of recording, which limits its application. The fragile watermark methods allow can be applied within strict and complete service system
Digital signature	A private key is used to encrypt a hashed version of the image. An associated public key can be used to decrypt the signature. The image can be hashed using the original hashing function	Digital signature cannot be used to authenticate a compressed image
Source recognition	Source recognition relies on the underlying characteristics of the components of digital devices. Support vector machine is usually used to identify device classes	Many manufacturers use the same components (i.e. lens, size of the sensor, choice of CFA). Therefore, the discriminatory power of some source recognition techniques are limited
Authenticity detection	Authenticity detection techniques for image forensics operate in absence of watermark or signature. The limitation of authenticity detection is minimal in all forensic scenarios	The accuracy of authenticity detection methods is still unsatisfactory

Source: Author.

The highlight and weakness of different types of forensic methods can be summarized in Table 1.

2.3 Major type forgery detections

In this sector, we brief two major types of forgery detections: clone detection and splicing detection.

2.3.1 Clone detection

Clone is the most common image forgery, which usually hides certain details or duplicates certain blocks of an image. Clone detection methods usually capture the inconsistence of color and noise variation of the target regions.

Blurring is usually used along the border of the modified blocks to lessen inconsistence. Most clone detection methods examine each pair of parts subdivided from the image. In order to reduce the processing time and enhance the robustness of the examine, some techniques use representations for dimensionality reduction, for example Discrete Cosine Transform (DCT), Principal Component Analysis (PCA).

In [13], Fridrich et al. divided an image into overlapping blocks of equal size firstly. Coefficients of each blocks was extracted by DCT. Afterwards, the duplicated blocks were detected by matching the quantized coefficients which had been lexicographically sorted. Best balance between performance and complexity was obtained using blocks matching algorithm. Popescu and Farid [14] extracted the coefficients by PCA instead of DCT. PCA-based detection can reduce the computational cost: the number of computations required are $O(N^t N \log N)$, where N^t is the dimensionality of the truncated PCA representation and N is the number of image pixels. Average detection accuracies obtained was 50% when JPEG quality is 95 with block size of 32×32 and 100% when JPEG quality is 95 with block size of 160×160. Accuracy degrades for small block sizes and low JPEG qualities. Since the dimensions of the coefficients extracted by PCA are smaller than that of DCT, [14] is demonstrated to be more effective than [13]. The weakness of [14] is that it cannot detect the rotating copy regions.

The essential of block-based methods is to compare blocks in an efficient manner, and invariant to some transformations through an appropriate choice of the feature representation. Fig. 8 presents a common framework of clone forgery detection.

Although block-based approaches seem effective to detect duplicated regions, they are incompetent to deal with the forged image which may manipulated by multiple kinds of transformations, such as scaling, rotation, flipping, blurring, illumination change and multiple cloning [15]. Unlike block-based techniques, keypoint-based methods identify and select keypoints in the image, instead of blocks. A feature vector is extracted per keypoint. In the field of computer vision, local visual features (e.g. Scale Invariant Feature Transform (SIFT) [16], Speeded-Up Robust Features (SURF) [17]) have been widely used for image retrieval and object recognition. They are regarded as invariance to numbers of geometrical transformations, and quite suitable for detecting the duplicated regions which were distorted.

In [18,19], a SIFT technique [16], which invariant to various geometrical transformations, is used to extract the features. Forgery detection is

Fig. 8 A common framework of block-based clone forgery detection. *Source: Author.*

performed while a number of SIFT features are matched. In [20], SURF features [17] is extracted instead of SIFT. However, the detection result is not obviously improved since the transformation invariance of SURF is little more than SIFT [21]. Transform–invariant features are obtained from the

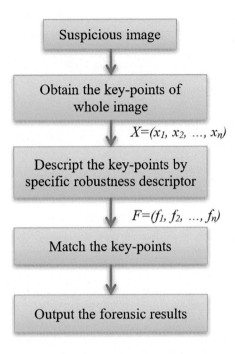

Fig. 9 A common framework of feature-based clone forgery detection. *Source: Author.*

MPEG-7 image signature tools in [22]. Such forgery detection approach obtains a feature matching accuracy in excess of 90% across postprocessing operations and are able to detect the cloned regions with a high true positive rate and lower false positive rate. As shown in Fig. 9, a common clone detection method mostly includes three main steps.

2.3.2 Splicing detection

Image splicing is another common type of forgery. Most splicing forgery detection tools make use of clues of abnormal in statistical characteristics. Support Vector Machine (SVM) is usually used in classifying forensic from suspicious images. A flowchart of splicing detection is shown in Fig. 10.

Popescu and Farid [14] used Expectation-Maximization (EM) algorithm to generate a probability map (p-map) from the residue of a local linear predictor to expose periodicities introduced by interpolation and

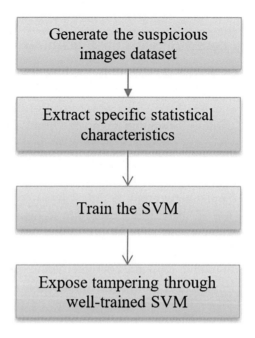

Fig. 10 A flowchart of splicing forgery detection. *Source: Author.*

resampling. However, the EM-based method is vulnerable to JPEG attacks. The periodic JPEG blocking artifacts would interfere with the periodic patterns by resampling.

Mahdian and Saic [23] proposed a blind and automatic approach to find the traces of resampling and interpolation. Wei et al. [24] discovered that there are peaks at several composite frequencies related to the parameters of both the first and the second operations. However, their work remains uncertain as to whether characteristic spectral peaks are appeared in all transformations. Furthermore, their work is limited to 1D case.

Feng et al. [25] firstly calculate the normalized energy density present within windows of varying size in the second derivative of the image in the frequency domain. The characteristic to derive a 19D feature vector is generated, and is used to train a SVM classifier. Ryu and Lee [26] proposed a technique to detect resampling on JPEG compressed images. They added noise before passing the image through the resampling detector. Most existing methods manually extract reliable features, and then feed them into a classifier (such as SVM). The classifier needs be trained with lots of labeled images for detection [27].

3. Forgery detection with deep learning

In computer vision, Deep Learning has demonstrated outstanding strength in different visual recognition tasks, such as image classification and semantic segmentation. Image forensic is to resolve classification problem. Many Deep Learning communities employ Deep Learning technique based on the development of multi-level Artificial Neural Network (ANN). Compared to Shallow Learning [28] (for example, SVM), the architecture of Deep Learning contains more hidden layer as presented in Fig. 11.

Convolutional Neural Network (CNN) first appeared in the late 1980s, as an extended version of ANN. It was applied in recognizing handwritten zip code. As the number of parameters in CNN is less than that in fully connected models and intuitive structure, CNN has become one of the most popular architectures used in resolving computer vision problems [29]. It is able to adaptively learn classification features instead of relying on hand-designed features. Therefore, we employ CNN to construct our forensic scheme.

The construction process will include: firstly, developing an efficient architectural of neural network; then, this artificial neural network would be trained by huge number of tampered and authentic image. Lastly, the features of faked image can be learned by the well-trained network, and the ability to identify the faked image from original ones can be obtained.

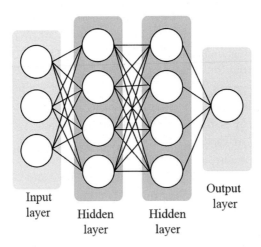

Fig. 11 Example of the architecture of Deep Learning. *Source: Author.*

As pointed out in Section 2.1, there are various type of forgery threats, and in many cases, additional transformations are used in order to cover forgery of the image. Pre-operations such as scaling and rotation, and post-operations such as filtering, are usually applied to make the forged regions more consistent with the whole image. Therefore, in order to combat the complicated threats, a universe approach, which is able to tackle varies types of forgery, is more needed than a single-function one. As shown in Fig. 10, many existing passive image forgery detection approaches extract features from images first, then select a classifier and train the classifier using the features extracted from training image sets, and finally classify the features. Deep Learning can more accurately discover specific features which can be leaned accurately by artificial neural network than traditional forensic approaches. In this sector, we brief development forensic based on Deep Learning.

3.1 Convolutional Neural Network in image classification

A CNN is a special type of multilayer neural network used in Deep Learning. It has recently drawing big attention in the computer vision fields. CNN is well-known for robustness to small inputs variations, minimal preprocessing and no requirement for any specific feature extractor choice. The CNN architecture relies on several deep neural networks that alternate convolutional and pooling layers. These deep neural networks belong to a wide class of models generally termed multi-stage architectures, which can be traced to Hubel and Wiesel's 1962 work on the cat's primary visual cortex [30]. Instead of relying on hand-designed features, CNNs are able to adaptively learn classification features. CNN has recently been proved to be effective solutions for problems such as object recognition, object tracing, and cell classification. These recent advances have been advanced by the use of GPUs to overcome the computational expense of estimating the large number of hyper-parameters that a deep network involves [31].

CNN is able to adaptively learn some invisible (or unreadable) features rather than human-selected features. These features are extracted by several set of convolutional layers. The coefficients of layers in each level are shared. In a classical CNN, convolution layer is followed by subsampling layer which is used to reduce the feature maps size. This subsampling layer is name as pooling layer. Fig. 12 demonstrates a classical CNN architecture. It contains two convolutional layers, two pooling layers and two full connection layers.

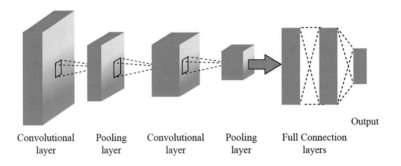

Convolutional   Pooling   Convolutional   Pooling   Full Connection
layer   layer   layer   layer   layers

Output

Fig. 12 A classical CNN architecture. *Source: Author.*

While the particular design or architecture of CNN may vary, they are built using a common set of basic elements and share a similar overall structure. The first layer is a convolutional layer, comprising several convolutional filters applied to the image in parallel. These filters act as a set of feature extractors. The convolutional outputs from all inputs are then transferred into element-wise nonlinearity. Pooling layer can reduce the spatial resolution of each feature map and translates information into more global one. Max-pooling technique can lead to faster convergence and improved generalization [32]. Via alternating convolutional layer and the output feature vectors are fed into the classification layer. The final layer in a classic CNN is a classification layer which output the probability of one sample classified into each class through Softmax connection [33]. Softmax is generated from logistic regression for multi-classification problems. Since it is easy to perform, it has been widely used in practical applications, such as digital image classification.

3.2 Literature review

Deep Learning in detecting forgery has drawn attention from many researchers recently [31,34,35].

Certain local structural relationships existed in the pixels would usually be independent to the image content. Tempering operations would alter these local relationships in a detectable way. Therefore, detection feature extractors must learn the relationship between a pixel and its local neighborhood while simultaneously suppressing the content of the image so that content-dependent features are not learned [31]. However, traditional CNN learn features of an image's content, rather than features for forgery detection. To overcome this problem, Bayar et al. [31] developed a new form of convolutional layer that is designed to suppress an image's content and adaptively learn forgery detection features. Prediction error filters are used to predict the pixel value at the center of the filter window. Then, the central value was subtracted to produce the prediction error. Experiment results [31] confirmed the efficiency of their methods. However, the boundary feature in manipulated region is usually affected by the texture in natural image. Furthermore, some forgery do not exhibit clear boundary discrepancy after the first convolution. Thus, the overall accuracy of Bappy et al. [36] is unacceptable.

In order to formulate the framework, Bappy et al. [36] employ a hybrid CNN and Long Short-Term Memory (LSTM) hybrid model to capture discriminative features between tampered and non-tampered regions. One of the key properties of manipulated regions is that they exhibit discriminative features in boundaries shared with neighboring non-tampered pixels. The aim of their approach is to learn the boundary discrepancy between tampered and non- tampered areas with the combination of LSTM and convolution layers. They developed an end-to-end training network to learn the parameters through back-propagation. The output of their network is the ground truth mask information. The framework is capable of detecting different types of image manipulations, including copy-move, removal and splicing. But the accuracy of their method is still satisfactory in practice.

To further improve the accuracy of forensic, Bunk et al. [34] proposed two methods to detect image forgery based on a combination of resampling features and Deep Learning. In the first method, the Radon transform of resampling features are computed on overlapping image patches. Deep Learning classifiers and a Gaussian conditional random field model are then used to create a heatmap. Tampered areas are identified using Random Walker Segmentation (RWS) function. In their CNN–LSTM method, resampling features computed on overlapping image patches are passed through CNN–based network for classification and localization. This system used a new convolutional layer as the first layer of an end-to-end patch classification network to learn to extract resampling features itself, rather than using the hand-crafted resampling features described before as step of their pipeline. However, the method does not function with images which have been degraded by JPEG depression.

Analysis of sensor pattern noise signatures, and in particular Photo-Response Non-Uniformity (PRNU), is one of the most powerful forensic techniques for digital photographs. Imperfections of imaging sensors introduce consistent noise, characteristic for each device, which enables reasoning about the origin and authenticity of photographs. Identification of signature inconsistencies in various regions of an image leads to a localization map indicating the most likely tampered content and is invaluable for discovering intentions of a forger. To detect such abnormal PRNU, Korus and Huang [35] proposed a multi-scale network model. Firstly, they consider a multi-scale fusion approach which involves combination of multiple candidate tampering probability maps into a single, more reliable decision map. The candidate maps are obtained with sliding windows of various sizes and thus allow to exploit the benefits of both small and large-scale analysis. Subsequently, they extend this approach by introducing modulated threshold drift and content-dependent neighborhood interactions, leading to improved identification performance with superior shape representation and easier detection of small forgeries. However, this compression-based method does not function on detecting fake colored image. Furthermore, some forgery operation may not affect the PRNU of an image, which would lead to a poor detection result.

In order to overcome the shortcomings of the systems in the above systems, we develop a new CNN system to tackle the detection challenges. Studies in relate works [31,33] had indicated that the original images cannot be directly fed into conventional CNN in digital forensic, as the forgeries tend conceal the traces not only visually but also statistically. Therefore, we develop the micro neural networks to extract the forensic features.

3.3 Generation of a multichannel feature map

Some common inherent statistics in frequency domain cannot be preserved well. A Multichannel Feature Map (MFM), which containing the periodicity property and filtering residual feature, is developed as the input of CNN to expose the inherent statistics. By using the MFM, an image is divided into different features in different channel which could be efficiently learned by different sub-network. Because our method uses two types of features, i.e. periodicity of interpolated signals and Filtering Residual Feature Map. Two sets of feature vectors constitute MFM, and are feed into neural network for training.

3.3.1 Hidden periodicity of interpolated signals

More geometric transformations are usually performed when creating a fake image. These geometric transformations typically involve resampling (e.g. scaling or rotating) which in turn calls for interpolation (e.g. nearest neighbor, bilinear). Specific statistical changes due to interpolation step can be identified as possible image forgery [37].

Geometric transformations consist two steps:

1) A spatial transformation of the physical rearrangement of pixels in the image is performed.

In the first step, coordinate transformation is described by a transformation function. The coordinates of the input image pixel are matched to the point in the output image:

$$x' = T_x(x, y) \quad y' = T_y(x, y) \tag{1}$$

General affine transformation can be described by the following equations:

$$x' = t_1 + t_2 + t_3 y \quad y' = l_1 + l_2 x + l_3 y \tag{2}$$

2) Pixels intensity values of the transformed image are assigned using a constructed low-pass interpolation filter $f(x)$.

The filter fk is multiplied with the proper filter weights when convolving them with w to compute signal values at arbitrary locations in the second step.

The spectrum does not overlap in the Fourier domain when the Nyquist criterion is satisfied. By using the optimal sinc interpolator the original signal can be well reconstructed from its samples. The sinc function is hard to implement in practice because of its infinite extent.

3.4 Periodicity property of forensic

The periodicity property exposes the trace of affine transformation. However, many forgeries are performed without affine transformations. The specific periodic peaks are hard to be detect in this case. On the other hand, as a popular noise removal and image enhancement tool, smooth filtering is applied widely for blurring and de-noising. In general, smooth filtering operations fall into two categories. One is linear, mainly including average and Gaussian filtering, and the other is nonlinear, i.e. median filtering.

According to Wei et al. [24], a signal $s(x)$ is sampled with a step size $\Delta \in R^+$ to produce a discrete data sequence $s_m = s(m\Delta)$. The signal can be reconstructed from its samples:

$$s^{(\varphi)}(x) = \sum_{m=-\infty}^{\infty} s_m \varphi\left(\frac{x}{\Delta} - m\right) \tag{3}$$

where $\varphi(\cdot)$ is the interpolation function. For linear interpolation $\varphi(x) = 1 - |x|$ where $|x| \leq 1$. Mahdian and Saic [23] had shown that the interpolation would bring into the signal and their derivatives a specific periodicity. And the periodicity is dependent on the interpolation kernel used. They generalized the method to the kth order derivative as:

$$D^{(k)} s^{(\varphi)}(x) = \sum_{m=-\infty}^{\infty} s_m D^{(k)} \varphi\left(\frac{x}{\Delta} - m\right) \tag{4}$$

where $D^{(k)}$ is an operator of the kth order derivative:

$$D^{(k)} s^{(\varphi)}(x) = \begin{cases} s^{(\varphi)}(x), k = 0 \\ \dfrac{\partial^k s^{(\varphi)}(x)}{\partial x^k}, k > 0 \end{cases} \tag{5}$$

In discrete signals derivative is typically approximated by computing the finite difference between adjacent samples. The variance of $D^{(k)} s^{(\varphi)}(x)$ is periodic over x with period Δ:

$$var\left\{D^{(k)} s^{(\varphi)}(x)\right\} = var\left\{D^{(k)} s^{(\varphi)}(x + \vartheta\Delta)\right\}, \vartheta \in \mathbb{Z} \tag{6}$$

Then, the 1D model can be analogously extended to the 2D case

$$var\left\{D^{(k)} s^{(\varphi)}(x, y)\right\} = var\left\{D^{(k)} s^{(\varphi)}(x + \vartheta\Delta_x, y + \vartheta\Delta_y)\right\}, \vartheta \in \mathbb{Z} \tag{7}$$

where $var\{\cdot\}$ means variance. In this paper, we will confine to the case of $k = 2$. The derivative kernel is $[1, -2.1]$. The periodicity is dependent on the interpolation kernel used.

The spatial transformation of an original image $f(x_1, x_2)$ maps the intensity value at each pixel location (x_1, x_2) to another location (y_1, y_2) in the new image $g(y_1, y_2)$. The most commonly used is the affine transformation that combines several linear operations like translation, rotation, scaling, skewing, etc. The mapping can be expressed as:

$$g = Af \tag{8}$$

where A is the 2×2 matrix that defines the linear transformation. In general, the pixels in the resulting image will not map to exact integer coordinates on the source image, but rather to intermediate locations between source pixels. Thus, the pixel interpolation algorithm would be used while performing the most spatial transformations [38]. The interpolation can be written as:

$$s(Af') = \sum_{f \in \mathbb{Z}^2} s(f)h(Af' - f) \tag{9}$$

where $s(\cdot)$ is the signal of interest and $h(\cdot):\mathbb{R}^2 \to \mathbb{R}^2$ is the interpolation kernel. For our theoretical analysis we will assume $s(x)$ to be a wide sense stationary signal. Many different interpolation filters are available with different characteristics, but the most common are the nearest neighbor, linear, cubic and truncated sinc.

The periodic properties of interpolation in special domain have been analyzed in several previous study [23,39]. In our scheme, the frequency feature map is used as the input of the CNN. Therefore, we focus on identifying the periodicity properties in frequency domain. The analyzation is basically originates from [40]. Taking the Fourier transform of $s(x)$ of the second order derivative, we have:

$$s(f) = \sigma^2 \, Cov(Th(f), Th(f))\omega(f) \tag{10}$$

where $f \in \mathbb{R}^2$ is the frequency. $Th(f)$ is the Fourier transform of the interpolation kernel $h(x)$ and $\omega(f)$ is the 2D Dirac comb function. The spectral lines are related to the transformation matrix A in a similar manner. Let det A and A^T denote the determinant and the transpose of A, respectively. The warped version of $s(f)$ in frequency domain can be expressed as:

$$S(f) = |\det A| \sum_{i \in \mathbb{Z}^2} s(A^T(f - i)) \tag{11}$$

Similar to [40], the baseband spectrum is represented in the range of $[0,1)^2$. The periodic peaks appear at normalized frequencies $\widetilde{f}_A^{(m)}$:

$$\widetilde{f}_A^{(m)} = frac\left\{(A^T)^{-1}m\right\}, m \in \mathbb{Z}^2 \tag{12}$$

where $frac\{\cdot\}$ denotes the element-wise fractional part sawtooth function. The periodic peaks are confirmed in the affine transformed region, and the periodicity property could be automatically learned by the CNN.

3.5 Filtering residual feature map

Many forgeries may not perform affine transformation. The specific periodic peaks are hard to be detected in such case. Therefore, we expend our scheme to include additional post-processed manipulations detection (e.g. filtering detection). Image filtering is a common post-process operation which may be exploited to reduce the discontinuity at the border of the forged object. The filtering process can be defined as:

$$g(i,j) = f(i,j) \times F_w(i,j)$$

$$= \sum_{k=-n}^{n} \sum_{l=-n}^{n} f(i-k, j-l) \times G_w(k,l), w = 2n+1 \qquad (13)$$

where $g(i,j)$ and $f(i,j)$ is the filtered and unaltered image, respectively. $F_w(i,j)$ is the filter function with the filter window of $w \times w$.

There are few perceptible differences between the original and the filtered image. The residual feature of filtering can hardly be learned by CNN. We then explore the feasibility of detecting the forgery in frequency domain. Inspired by the prominent results in median filtering residual forensic, we calculate the difference of between original image and its filtered version in frequency domain. The Filtering Residual in Frequency (FRF) features of an image is defined as:

$$FRF(i,j) = FFT2(f_w(X(i,j))) - FFT2(X(i,j)) \qquad (14)$$

where FFT2 is 2D Fast Fourier Transform (FFT), f_w is the filtering process with the a $w \times w$ filter window. We translate the spectrum and take logarithm transform to the FRF value to improve the identification. The transformation can be expressed as:

$$F' = log(shift(F) + 1) \qquad (15)$$

where F and F' is the original and transformed spectrum, respectively. Then (14) can be written as:

$$FRF = log(shift(FFT2(f_w(I))) + 1) - log(shift(FFT2(I)) + 1) \qquad (16)$$

To illustrate the generation of FRF features clearly, we frame the filtering detection as differentiating between the following two hypotheses:

H_u: $X(i,j)$ is not an unaltered image.

H_a: $X(i,j)$ is an altered image, i.e. $X(i,j)$ has been smooth filtered.

Under hypothesis H_a, $X(i,j)$ is filtered by f_u with filter window $u \times u$, the FRF can be presented as:

$$FRF'(i,j) = FFT2(f_w(f_u(I(i,j)))) - FFT2(f_u(I(i,j))) \qquad (17)$$

where $I(i,j)$ is the original unaltered image. The residual between original image and its filtered version can be presented as a band-pass filter signal. Under hypothesis H_u, $X(i,j)$ is unfiltered, the FRF values are still obtained

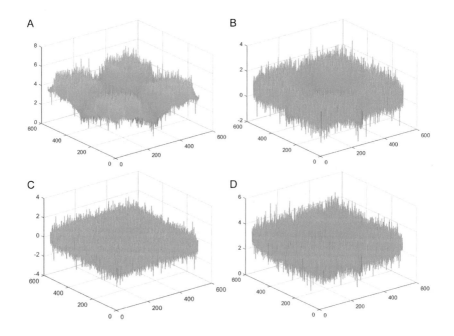

Fig. 13 The FRF of (A) average filtering, (B) Gaussian filtering, (C) median filtering, and (D) bilateral filtering.

by (14). If an unaltered image has been filtered twice, the parameters of band–pass filter signal would be different [42].

Thus, the filtering detection task is to identify the difference between $FRF(i,j)$ and $FRF'(i,j)$. Fig. 13 presents the filtering residual in frequency domain of four types of filtering. The FRF images exhibit distinct patterns which could be learned by CNN.

4. Our scheme: Toward a universal system of forgery detection

As the convolutional layers extract features of an image's content, we directly fed the raw image pixels into the conventional CNN models. Since many tampered images do not contain any visual clue of alteration, the classification result was about 50%, which means that the network can barely learn from the input images. To increase the performance in forgery detection, an additional layer is proposed in our multichannel CNN architecture. As shown in Fig. 14, our CNN scheme contains a f-maps layer, a micro neural networks module, a convolutional module and a linear classification

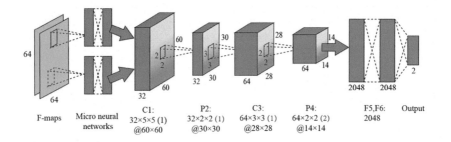

Fig. 14 Architecture of our CNN scheme. *Source: Author.*

module (composed of two fully connected layers and a softmax layer). Each layer (except the micro neural networks) is denoted by index *l*.

4.1 Micro neural networks

Deep convolutional neural network has recently been applied to image classification with large image datasets. A deep CNN is able to learn basic filters automatically and combine them hierarchically to enable the description of latent concepts for pattern recognition. However, many deep CNNs have the problems of overfitting and huge processing time [41].

Most data are not separable by linear filters since they are usually distributed on nonlinear manifolds. To enhance model discriminability, the Network In Network (NIN) model was developed by [42]. Multilayer Perceptron (MLP) was created by using a sliding micro neural network to increase the nonlinearity of local patches. The NIN structure can efficiently abstract the quantities of information within the receptive fields.

Kim et al. [43] proposed a NIN based network module that can be considered as a modification of inception structure. In [43], the pool projection and pooling layer are removed for maintaining the entire feature map size, and a larger kernel filter is added instead. The number of parameters on account of removed dense prediction and pooling is greatly reduced. They added a larger kernel than the original inception structure for not increasing the depth of layers. Their structure is applied to typical image-to-image learning problems, i.e. the problems where the size of input and output are same such as skin detection, semantic segmentation, and compression artifacts reduction. The NIN based module is shown in Fig. 15. Their architecture is consisting of four convolution layers with nonlinear layers and eight modified inception modules. Input image goes through the convolutional network to be a skin probability map at the output.

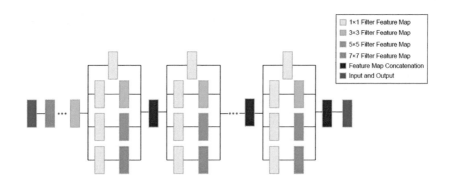

Fig. 15 Network-in-network structure. *Source: Y. Kim, I. Hwang, N.I. Cho, A New Convolutional Network-in-Network Structure and Its Applications in Skin Detection, Semantic Segmentation, and Artifact Reduction (2017).*

In many studies [33,44–49], the features came from the feature extraction step need to be consolidated before the network training. We use a different strategy to skip the feature fusion step. The features are separately characterized in the neural networks. We establish two micro neural networks proposed in NIN structure to abstract the data within the receptive field. The feature map can be calculated as follows:

$$f_{i,j,m} = \max\left(w_m^T x_{i,j}, 0\right) \tag{18}$$

where (i, j) is the pixel index in the f-map, $x_{i,j}$ stands for the input patch cantered at location (i, j), w is the weight and m is used to index the channels of the f-maps. Then, the calculation performed by micro neural network is shown as follows:

$$f_{i,j,m}^1 = \max\left(w_{m_n}^1{}^T x_{i,j} + b_{m_1}, 0\right) \tag{19}$$

$$f_{i,j,m}^n = \max\left(w_{m_n}^n{}^T f_{i,j,m}^{n-1} + b_{m_n}, 0\right)$$

where n is the number of layers in the multilayer perceptron, b_{m_n} is the bias of nth layer in mth channel. Rectified linear unit is used as the activation function in the multilayer perceptron.

Each micro neural network is an MLP consisting of multiple fully connected layers with nonlinear activation functions. The MLP is shared among all local receptive fields and without the Softmax layer. Unlike the structure in [42], we divide the input feature maps into two f-maps, as shown in Fig. 16. One f-map is for periodicity property abstraction, and another is

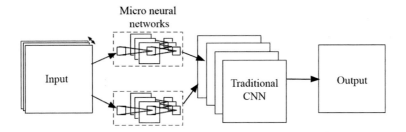

Fig. 16 Micro neural networks: Our neural network structure. *Source: Author.*

for FRF abstraction. The proposed micro neural network consists two fully connected layers. Each fully connected layer has 1024 neurons.

4.2 Generation of f-map layer

The traces left by forgery can be investigated after suppressing the interference of irrelevant information (e.g. image edges and textures). The f-map layer is essential in our CNN model. In such layer, the network will take an image patch as input, and output the multichannel feature map. There is a trade-off in selecting the size of the image patch: forged region is more detectable in larger patch sizes because the filter signal is more distinct, but small forged area will not be localized that well. Finally, we choose 64×64 pixels as the size that we can detect reasonably well.

We calculate the periodicity property of each image patch by using formula (14). And the filtering residual features in frequency domain are simultaneously calculated by using formulas (16) and (17). The residual between original image and its filtered version can be presented as a band-pass filter signal. Then the f-map layer which consisted of two types of feature is obtained. The flowchart of the generation of f-map layer is shown in Fig. 17.

4.3 Convolutional module

The convolutional layers are capable of extracting different features from an image such as edges, textures, objects, and scenes [50]. As pointed above, forgery is better captured around the boundary of forgery regions. Thus, the low-level features are critical to identify manipulated regions. The filters in convolutional layer will create feature maps that are connected to the local region of the previous layer.

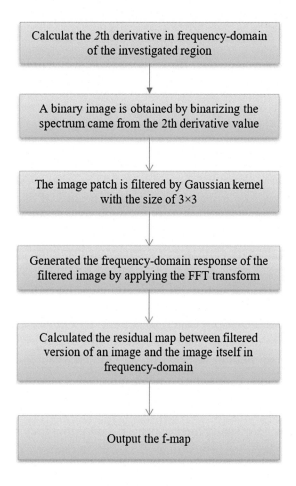

Fig. 17 The flowchart of generation of f-map. *Source: Author.*

Two pairs of convolutional (C1 and C3) and pooling layers (P2 and P4) are designed following the micro neural network in our CNN. In the convolutional layers, we use kernel size of $m \times m \times C$, where C is the depth of a filter and m is the size of convolutional kernel. The parameters C and m have different values for different layers in the network as is demonstrated in Fig. 14. For example, the convolutional kernel for the first layer is $5 \times 5 \times 32$. The size of the output (C1) is $64 \times 64 \times 32$, which means the number of feature maps is 32 and the resolution of feature maps is 64×64. The convolution operation can be denoted as:

$$x^l_j = \sum_{l=1}^{n} x_i^{l-1} \times k_{ij}^{l-1} + b^l_j \tag{20}$$

where $\times$ denotes convolution, x^l_j is the jth output map in layer l, the convolutional kernel k_{ij}^{l-1} (also called weight) can be updated while training the network. It connecting the ith output map in layer $l-1$ and the jth output map in layer l. b^l_j is the trainable bias parameter of the jth output map in layer l.

The pooling layer is used for a down sampling operation after obtaining feature maps through convolution process. In classical CNN, convolution layers are followed by a subsampling layer. The size of effective maps is reduced by pooling layer, and some invariance features are introduced. A max-pooling layer is a variant which has shown some merit in [51]. The output of a max-pooling layer is given by the maximum activation over non-overlapping regions, instead of averaging the inputs as in a classical subsampling layer. A bias is added to the resulting pooling and the output map is passed through the squashing function.

In our network, a max-pooling layer with filter of size 2×2 is used to decrease the size of feature maps to 30×30 after C1 layer. Let l denotes the index of a max-pooling layer. The layer's output is a set P_l of square maps with size w_l. We get the P_l from P_{l-1}. The square maps size w_l is obtained by $w_l = w_{l-1}/k$, where k is the size of the square max-pooling kernel. Following the pooling layer (P2) is another pair of convolution and pooling layer with 64 kernels of size 3×3 and a filter of size 2×2. Dropout [52] is a wildly used technique for avoiding overfitting in neural networks. Therefore, the Rectified Linear Units (ReLUs) [53] and dropout are used in our proposed CNN architecture. Based on Eq. (17), the operation is expressed as:

$$f_{m,n} = \max\left(x^l_{m,n}, 0\right) \tag{21}$$

where $x_{m,n}$ is for the input patch centered in the feature map point (m, n) in layer l.

4.4 Classification module

In our scheme, the classification module is placed after the convolutional module. Specific parameters of convolution and pooling layers are selected. A classification module is used to classify the features that came from the former layers into different types. It usually consists fully connected layers and a Softmax layer. In Fig. 18, a classical classification module has been presented.

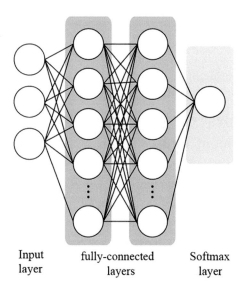

Input   fully-connected   Softmax
layer   layers   layer

Fig. 18 A classical classification module. *Source: Author.*

The output feature maps of the last convolutional layer are then down-sampled to 1 pixel for each feature map. A one-dimension vector of parameters is obtained. Finally, classical feed-forward fully connected layers are used to perform the classification task.

Softmax model is the extension of logistic regression model on multiple classification problems, in which the number of categories to be classified is greater than 2 and the categories are mutually exclusive [34]. Softmax is typically used in the last layer of the neural network for final classification and normalization. A Softmax function can be identified as:

$$f\left(z_j\right) = \frac{\exp\left(z_j\right)}{\sum_{i=1}^{n} \exp\left(z_i\right)}$$

where z_i is the input of last convolutional layer.

The last layer, in the case of supervised learning, contains as many neurons as the number of classes. Our classification module consists two fully connected layers and followed by a two-way Softmax loss layer. Each fully connected layer (F5 and F6 in Fig. 14) has 2048 neurons. Dropout is used in both fully connected layers. The output of the last fully connected layer has two neurons which is fed into Softmax. We use back-propagation algorithm to train our CNN. The weights and the bias in the convolutional and fully connected layers are updated adaptively by performing the error propagation procedure. Finally, we feed back the classification results to guide the feature extraction automatically.

5. Experiment results

5.1 Experiment setup

We conduct experiments on our scheme on a computer with one GPU (Nvidia GeForce GTX 1080Ti with 11GB RAM). To evaluate the performance of the proposed model and compare its performance with other schemes, we use a composite image database containing 12,120 images for testing. These images are from three image databases: the BOSSbase 1.01 [54], CASIA v2.0 [55], and laboratory database. From BOSSbase, we use 10,000 uncompressed images with the size of 512×512. From the rest databases, we use 2120 images with the resolution of 348×256 to 4032×3024. All images are firstly converted into grayscale. Then we randomly select 70% images as the training set, while the other 30% as the testing set. Each image in training set was cropped to 64×64 region from the center. The training and the testing set were generated as follow strategy. Theoretically, filtering operations on images with any parameter could be implemented. But in practice, only a limited number of typical parameters are applied. As exhibited in Table 2, we applied five types of operations with different factors to every image.

In total, we create a training set containing 178,164 altered blocks and 8484 unaltered blocks. To measure the detection performance, True

Table 2 Five types of operations.

Operation	Factor/Parameter
Scaling	Scaling factor is in the range of $\{0.5, 0.7, 0.9, 1.1, 1.3, 1.5, 1.7, 1.9\}$
Rotating	Rotation angle is in the range of $\{1°, 2.5°, 5°, 15°, 30°, 45°, 80°\}$
Median filtering	Size of the kernel is $\{3 \times 3, 5 \times 5\}$
Gaussian filtering	Size of the kernel is $\{3 \times 3, 5 \times 5\}$
Average filtering	Size of the kernel is $\{3 \times 3, 5 \times 5\}$

Source: Author.

Positive Rate (TPR) and False Positive Rate (FPR) were used. Where TP and TN denote as the number of the true detection of forged images and original images, respectively. Denote the FP and FN as the number of the wrong detection of original images and forged images, respectively. Then, the TPR is the fraction of tampered images correctly identified as such, while FPR is the fraction of identifying an original image as a tampered one. They can be represented as:

$$\text{TPR} = {^{\text{TP}}}\!/\!_{(\text{TP} + \text{FN})} \tag{22}$$

$$\text{FPR} = {^{\text{FP}}}\!/\!_{(\text{FP} + \text{FN})} \tag{23}$$

5.2 Experimental results

Table 3 summarizes the performance of our scheme in binary classification for detecting different image forgeries. Each operation is named by its first letter and its factor value, for example, S0.5 denotes scaling with factor of 0.5. Detection results demonstrate that our approach is able to handle most forgery.

We also compared our scheme to several other methods (Chang and Chen [41], Bappy et al. [36], Boureau [32] and Liu et al. [33]). SVM models for [32,33] were performed in MATLAB. The network models used in [36,41] were also generated in Caffe. Comparison results are presented in Table 4.

Table 3 TPR(%) and FPR(%) of different operations.

Operation	S0.5	S0.7	S0.9	S1.1	S1.3	S1.5	S1.7
TPR	93.23	93.52	92.55	92.09	93.87	94.65	95.32
FPR	8.33	9.23	8.88	8.16	9.85	7.99	7.65
Operation	S1.9	R1	R2.5	R5	R15	R30	R45
TPR	95.19	91.40	93.87	94.52	92.87	93.12	95.32
FPR	9.30	9.50	9.22	8.64	8.23	9.19	9.51
Operation	R80	M3	M5	G3	G5	A3	A5
TPR	95.79	93.09	94.19	94.75	95.74	94.12	94.77
FPR	8.92	7.66	8.31	7.87	8.11	7.53	8.33

Source: Author.

Table 4 TPR(%) and FPR(%) of different methods.

Method	Scaling	Rotating	Median filtering	Gaussian filtering	Average filtering
Ryu and Lee [26]	80.31	94.45	–	–	–
	15.42	8.87	–	–	–
Liu et al. [27]	–	–	91.41	92.45	90.75
	–	–	9.55	9.82	11.48
Xu et al. [56]	–	–	–	90.64	91.97
	–	–	–	10.42	11.2
Bappy et al. [36]	81.41	82.87	72.11	75.3	72.97
	13.54	15.2	15.56	15.22	14.62
Bayar and Stamm [31]	82.32	83.88	73.26	80.48	84.75
	11.49	9.72	9.81	10.55	9.49
Our scheme	93.8	93.84	93.64	95.25	94.44
	8.67	9.03	7.99	7.99	7.93

The first and the second row of each method is the TPR and FPR, respectively.
Source: Author.

Ryu and Lee [26] exploited the periodic properties of interpolation by the second derivative of the transformed image in both the row and column directions. They obtained a relative high performance in rotation detection. But the accuracy rate reduced when the target object scaled down. On the other hand, Liu et al. [27] and Xu et al. [56] proposed filtering forensic task. Although they obtained a remarkable score, but their method could not function well in detecting forgery without filtering.

As demonstrated in Table 3, the methods of Bappy et al. [36] and Bayar and Stamm [31] are both universal forgery detection approaches. Bappy et al. [36] proposed a LSTM network to exam the boundary discrepancy around the tampered region. They discovered that the boundaries between manipulated and neighbored non–manipulated regions exhibit discriminative features. However, the boundary feature in manipulated region is usually affected by the texture in natural image. Furthermore, some forgery do not exhibit clear boundary discrepancy after the first convolution. Thus, the overall accuracy of Bappy et al. [36] is unacceptable. Bayar and Stamm [31] proposed a new form of convolutional layer that will force the CNN to learn forgery detection features from images without requiring any preliminary feature extraction or pre-processing. The detection results of Xu et al. [56] is unacceptable since the new inserted layer unable directly abstract the features from a raw image. Table 3 demonstrates that our method outperforms these existing methods. Our CNN architecture can learn to distinguish if the image blocks had been tampered by using the f-map layer and obtains an average accuracy beyond 93%.

Fig. 19 presents two splicing detection examples in CASIA [55]. As can be seen, our method can perfectly identify the splicing regions. Fig. 20 presents two high resolution examples which had performed clone forensic. Note that the smooth operation was performed to blur the boundaries between manipulated and neighbored non-manipulated regions. The forensic attacks in Fig. 20 are probably the most difficult challenge to traditional clone forgery detection approaches. Because the erased regions were cloned and smooth several times, with no two blocks were the same in an image.

5.3 Robustness against JPEG attack

JPEG compression is a critical means to erase the trace of tampering. The above discussions for tampering detection are on JPEG compressed images only with a Quality Factor (QF) of 90. To estimate the performance of the proposed method in low quality images with strong JPEG compression. We summarize the TPR and FPR scores for all diverse JPEG quality factors with

Fig. 19 Examples of splicing detection. The detected blocks were marked in green.
Source: Author.

Fig. 20 Examples of clone detection. The detected blocks were marked in green. *Source: Author.*

range from 90 to 35. For comparison, two universal forensic approaches (Bappy et al. [36] and Bayar and Stamm [31]) were performed on the same testing set.

As demonstrated in Table 5, our scheme is robust to JPEG compression when the QF is greater than 50. The FPR is stable compare to the other two approaches. Our method does not function when QF of the input JPEG image is lower than 50.[a]

6. Summary of key contribution of the research

In real–life situations, image forgery detection should be able to detect all types of tampering rather than focusing on a specific type. Toward this goal, we propose a universal end-to-end scheme based on Deep Learning technique.

Our new scheme makes contributions as the follows:

---

[a] Few methods can generate good result in the condition where QF is lower than 50.

Table 5 Schemes' performances against JPEG compression.

Method	JPEG-90	JPEG-70	JPEG-50	JPEG-35
Bappy et al. [36]	78.11	78.23	67.61	48.96
	14.74	15.21	22.72	24.73
Bayar and Stamm [31]	82.32	81.42	73.56	62.87
	10.74	10.57	13.81	23.25
Proposed	94.28	93.89	86.94	80.1
	7.87	8.33	8.99	8.86

Source: Author; M.J.H. Bappy, A.K. Roy-Chowdhury, J. Bunk, L. Nataraj, B.S. Manjunath, Exploiting spatial structure for localizing manipulated image regions, in: International Conference on Computer Vision, 2017; B. Bayar, M.C. Stamm, A deep learning approach to universal image manipulation detection using a new convolutional layer, in: ACM Workshop on Information Hiding and Multimedia Security, 2016, 5–10.

➢ While most existing methods can only detect some specific forgery operation, our method is proposed to detect various pre-operations and post-operations.

➢ In traditional model, the convolutional layers extract features of an image, instead of learning filters that identify traces of the tampering. We developed an additional layer, in order to increase the performance in forgery detection.

➢ We establish two Micro Neural Networks to abstract data within the receptive field. The features are separately characterized in the neural networks. Thus, the traditional feature fusion step can be skipped.

➢ Our scheme uses image blocks to train the MCNN instead of using the whole image, so that the system can detect the splicing forgery and identify the forged area.

7. Research directions in the field

While Deep Learning emerges as a hot research direction in image forensic field, our study revealed some gaps in existing research outcomes. Image forensic should be able to detect almost all types of forgery rather than focusing on a particular type. The techniques that we reviewed represent important results for image forensic, especially considering that the problems they tackle were previously (almost) unexplored. A large set of tools is now available to investigate on image forensic. Despite of progress made by many researchers, big challenges continue to emerge in image forensic. For example, there is a pressing need to extend image forensic in video forensic. More and more forgery techniques are threatening the reliability of video, which can be divide into static pictures. Effective schemes for video forensic need to be established in future research.

8. Conclusion

In this chapter, we constructed a forensics-awareness scheme to support reliable image forensics. Unlike most existing forensic methods, which can only detect a certain type of image tampering, our system can capture various forgery operations, including clone, removal, splicing and smoothing.

While many CNN-based forensics systems have problems of requiring too much processing time, as data are not separable by linear filters since they are usually distributed on nonlinear manifolds. To increase the efficiency of extracting features from tampered image, we developed a novel Multichannel Neural Network. Through such network, our system is able to learn hierarchical representations from input images. The traces left by forgery are automatically learned by the improved neural network. Thus, tampered images can be identified using the well-trained network. Most existing system only detecting a certain manipulation, which may usually lead to misled results, e.g. results with irrelevant features and/or classifiers are used. To overcome this limitation, we extract the periodicity property and filtering residual feature from the image blocks. The f-map is then generated by linear blending the periodic spectrum and the residual map, and is fed into the network. The CPSSs with forensic-awareness is designed and tested on different datasets.

Experimental results confirm that our system has better performance than the formerly existing ones. The great potential of Deep Learning in image forensic has been revealed by our study, which will inspire more exploration in forensic methods from this perspective.

Acknowledgments

This research is supported by the Humanities and Social Sciences projects of the Ministry of Education (No. 18YJC760112); the Social Science Fund of Jiangsu Province (No. 18YSD002); and the Fundamental Research Funds for the Central University (No. 2019JDZD02).

Key terminology and definitions

Deep Learning Deep learning is part of a broader family of deep learning methods based on learning data representations, as opposed to task-specific algorithms. Learning can be supervised, semi-supervised or unsupervised.

Forensic Forensic technique collects, preserves, and analyses evidence during the course of an investigation. While some forensic experts travel to the scene of the crime to collect the evidence themselves, others occupy a laboratory role, performing analysis on objects brought to them by other individuals.

Watermark A watermark is an identifying image or pattern in paper that appears as various shades of lightness/darkness when viewed by transmitted light (or when viewed by reflected light, atop a dark background), caused by thickness or density variations in the paper. Watermarks have been used on postage stamps, currency, and other government documents to discourage counterfeiting.

Digital Signature A digital signature is a mathematical scheme for presenting the authenticity of digital messages or documents. A valid digital signature gives a recipient reason to believe that the message was created by a known sender (authentication), that the sender cannot deny having sent the message (non-repudiation), and that the message was not altered in transit (integrity).

Steganalysis The goal of steganalysis is to identify suspected packages, determine whether or not they have a payload encoded into them, and, if possible, recover that payload. Steganalysis generally starts with a pile of suspect data files, but little information about which of the files, if any, contain a payload.

Convolutional Neural Network (CNN) A special type of multilayer neural network used in deep learning. It has recently drawing great attention in the computer vision fields. CNN is well-known for robustness to small inputs variations, minimal pre-processing and do not require any specific feature extractor choice.

Scale Invariant Feature Transform (SIFT) SIFT is a feature detection algorithm in computer vision to detect and describe local features in images. SIFT keypoints of objects are first extracted from a set of reference images and stored in a database. An object is recognized in a new image by individually comparing each feature from the new image to this database and finding candidate matching features based on Euclidean distance of their feature vectors.

Speeded-Up Robust Features (SURF) SURF is a patented local feature detector and descriptor. It can be used for tasks such as object recognition, image registration, classification or 3D reconstruction. It is partly inspired by the scale-invariant feature transform (SIFT) descriptor. The standard version of SURF is several times faster than SIFT and claimed by its authors to be more robust against different image transformations than SIFT.

Support Vector Machine (SVM) SVM is supervised learning model with associated learning algorithms that analyze data used for classification and regression analysis. Given a set of training examples, each marked as belonging to one or the other of two categories, an SVM training algorithm builds a model that assigns new examples to one category or the other, making it a non-probabilistic binary linear classifier.

Fast Fourier Transform (FFT) FFT is an algorithm that samples a signal over a period of time (or space) and divides it into its frequency components. These components are single sinusoidal oscillations at distinct frequencies each with their own amplitude and phase. This transformation is illustrated in Diagram 1. Over the time period measured in the diagram, the signal contains three distinct dominant frequencies.

Multilayer Perceptron (MLP) MLP is a class of feed-forward artificial neural network. An MLP consists of, at least, three layers of nodes: an input layer, a hidden layer and an output layer. Except for the input nodes, each node is a neuron that uses a nonlinear activation function. MLP utilizes a supervised learning technique called back-propagation for training.

References

[1] D. Tan, L. Zhang, Q. Ai, An embedded self-adapting network service framework for networked manufacturing system, J. Intell. Manuf. (2016) 1–18.

[2] J. Yick, B. Mukherjee, D. Ghosal, Wireless sensor network survey, Comput. Netw. 52 (2008) 2292–2330.

[3] A. Juels, RFID security and privacy: a research survey, IEEE J. Sel. Areas Commun. 24 (2006) 381–394.

[4] G. Madlmayr, J. Langer, C. Kantner, J. Scharinger, NFC devices: security and privacy, in: International Conference on Availability, Reliability and Security, 2008, pp. 642–647.

[5] K. Khuspe, V. Mane, Robust image forgery localization and recognition in copy-move using bag of features and SVM, in: 2015 International Conference on Communication, Information & Computing Technology (ICCICT), IEEE, 2015, pp. 1–5.

[6] B. Yang, X. Sun, H. Guo, Z. Xia, X. Chen, A copy-move forgery detection method based on CMFD-SIFT, Multimed. Tools Appl. 77 (2018) 837–855.

[7] J. Wang, S. Lian, Y.Q. Shi, Hybrid multiplicative multi-watermarking in DWT domain, Multidim. Syst. Sign. Process. 28 (2015) 1–20.

[8] X. Chen, S. Chen, Y. Wu, Coverless information hiding method based on the chinese character encoding, J. Internet Technol. 18 (2017) 133–143.

[9] S. Etemad, M. Amirmazlaghani, A new multiplicative watermark detector in the contourlet domain using T location-scale distribution, Pattern Recogn. 77 (2018) 99–112.

[10] C.I. Podilchuk, E.J. Delp, Digital watermarking: algorithms and applications, IEEE Signal Process. Mag. 18 (2001) 33–46.

[11] D. Pointcheval, J. Stern, Security arguments for digital signatures and blind signatures, J. Cryptol. 13 (2000) 361–396.

[12] R. Anderson, S. Walter, B. Terrance, G. Siome, Vision of the unseen: current trends and challenges in digital image and video forensics, ACM Comput. Surv. 43 (2011) 1–42.

[13] J. Fridrich, B.D. Soukal, A.J. Lukáš, Detection of copy-move forgery in digital images, in: Proceedings of Digital Forensic Research Workshop, Cleveland, OH, 2003.

[14] A.C. Popescu, H. Farid, Exposing digital forgeries by detecting traces of resampling, IEEE Trans. Signal Process. 53 (2005) 758–767.

[15] V. Christlein, C. Riess, J. Jordan, E. Angelopoulou, An evaluation of popular copy-move forgery detection approaches, information forensics and security, IEEE Trans. Inf. Forensics Secur. 7 (2012) 1841–1854.

[16] D. Lowe, Distinctive image features from scale-invariant keypoints, Int. J. Comput. Vis. 60 (2004) 91–110.

[17] H. Bay, A. Ess, T. Tuytelaars, L. Van Gool, Speeded-up robust features (SURF), Comput. Vis. Image Underst. 110 (2008) 346–359.

[18] X. Pan, S. Lyu, Region duplication detection using image feature matching, IEEE Trans. Inf. Forensics Secur. 5 (2010) 857–867.

[19] I. Amerini, L. Ballan, R. Caldelli, A. Del Bimbo, G. Serra, A SIFT-based forensic method for copy-move attack detection and transformation recovery, information forensics and security, IEEE Trans. Inf. Forensics Secur. 6 (2011) 1099–1110.

[20] C. Neamtu, C. Barca, E. Achimescu, B. Gavriloaia, Exposing copy-move image tampering using forensic method based on SURF, in: 2013 International Conference on Electronics, Computers and Artificial Intelligence (ECAI), 2013, pp. 1–4.

[21] J. Luo, G. Oubong, A comparison of SIFT, PCA-SIFT and SURF, Int. J. Image Process. 3 (2009) 143–152.

[22] P. Kakar, N. Sudha, Exposing postprocessed copy-paste forgeries through transform-invariant features, IEEE Trans. Inf. Forensics Secur. 7 (2012) 1018–1028.

[23] B. Mahdian, S. Saic, Blind authentication using periodic properties of interpolation, IEEE Trans. Inf. Forensics Secur. 3 (2008) 529–538.

[24] W. Wei, S. Wang, X. Zhang, Z. Tang, Estimation of image rotation angle using interpolation-related spectral signatures with application to blind detection of image forgery, IEEE Trans. Inf. Forensics Secur. 5 (2010) 507–517.

[25] X. Feng, I.J. Cox, G. Doerr, Normalized energy density-based forensic detection of resampled images, IEEE Trans. Multimedia 14 (2012) 536–545.

[26] S.J. Ryu, H.K. Lee, Estimation of linear transformation by analyzing the periodicity of interpolation, Pattern Recogn. Lett. 36 (2014) 89–99.

[27] A. Liu, Z. Zhao, C. Zhang, Y. Su, Smooth filtering identification based on convolutional neural networks, Multimed. Tools Appl. (2016) 1–15.

[28] S. Argamon, I. Dagan, Y. Krymolowski, A memory-based approach to learning shallow natural language patterns, in: International Conference on Computational Linguistics, 1998, pp. 67–73.

[29] S. Li, Z.Q. Liu, A.B. Chan, Heterogeneous multi-task learning for human pose estimation with deep convolutional neural network, Int. J. Comput. Vis. 113 (2015) 19–36.

[30] D.H. Hubel, T.N. Wiesel, Receptive fields, binocular interaction and functional architecture in the cat's visual cortex, J. Physiol. 160 (1962) 106–154.

[31] B. Bayar, M.C. Stamm, A deep learning approach to universal image manipulation detection using a new convolutional layer, in: ACM Workshop on Information Hiding and Multimedia Security, 2016, pp. 5–10.

[32] Y.L. Boureau, F. Bach, Y. Lecun, J. Ponce, Learning mid-level features for recognition, in: Computer Vision and Pattern Recognition, 2010, pp. 2559–2566.

[33] J. Chen, X. Kang, Y. Liu, Z.J. Wang, Median filtering forensics based on convolutional neural networks, IEEE Signal Process. Lett. 22 (2015) 1849–1853.

[34] J. Bunk, J.H. Bappy, T.M. Mohammed, L. Nataraj, A. Flenner, B.S. Manjunath, S. Chandrasekaran, A.K. Roychowdhury, L. Peterson, Detection and localization of image forgeries using resampling features and deep learning, in: IEEE Computer Society Conference on Computer Vision and Pattern Recognition Workshops (CVPRW), 2017.

[35] P. Korus, J. Huang, Multi-scale analysis strategies in PRNU-based tampering localization, IEEE Trans. Inf. Forensics Secur. 12 (2017) 809–824.

[36] M.J.H. Bappy, A.K. Roy-Chowdhury, J. Bunk, L. Nataraj, B.S. Manjunath, Exploiting spatial structure for localizing manipulated image regions, in: International Conference on Computer Vision, 2017.

[37] G.K. Birajdar, V.H. Mankar, Digital image forgery detection using passive techniques: a survey, Digit. Investig. 10 (2013) 226–245.

[38] D. Va'Zquez-Pad'In, C. Mosquera, F. Pe'Rez-Gonza'Lez, Two-dimensional statistical test for the presence of almost cyclostationarity on images, in: IEEE International Conference on Image Processing, 119, 2010, pp. 1745–1748.

[39] M. Kirchner, Fast and reliable resampling detection by spectral analysis of fixed linear predictor residue, in: ACM Workshop on Multimedia and Security, 2008, pp. 11–20.

[40] C. Chen, J. Ni, Z. Shen, Y.Q. Shi, Blind forensics of successive geometric transformations in digital images using spectral method: theory and applications, IEEE Trans. Image Process. 26 (2017) 2811–2824.

[41] J.R. Chang, Y.S. Chen, Batch-Normalized Maxout Network in Network, Computer Science, 2015.

[42] M. Lin, Q. Chen, S. Yan, Network in Network, Computer Science, 2013.

[43] Y. Kim, I. Hwang, N.I. Cho, A New Convolutional Network-in-Network Structure and Its Applications in Skin Detection, Semantic Segmentation, and Artifact Reduction, arXiv, 2017.

[44] M. Kirchner, J. Fridrich, On detection of median filtering in digital images, in: Proc. SPIE, Electronic Imaging, Media Forensics and Security II, SPIE, 2010, pp. 1–12.

[45] G. Cao, Y. Zhao, R. Ni, L. Yu, H. Tian, Forensic detection of median filtering in digital images, in: IEEE International Conference on Multimedia and Expo, 2010, pp. 89–94.

[46] H.D. Yuan, Blind forensics of median filtering in digital images, IEEE Trans. Inf. Forensics Secur. 6 (2011) 1335–1345.

[47] J. Li, X. Li, B. Yang, X. Sun, Segmentation-based image copy-move forgery detection scheme, IEEE Trans. Inf. Forensics Secur. 10 (2015) 507–518.

[48] Z. Zhou, C.-N. Yang, B. Chen, X. Sun, Q. Liu, Q.M.J. Wu, Effective and efficient image copy detection with resistance to arbitrary rotation, IEICE Trans. Inf. Syst. E99-D (2016) 1531–1540.

[49] X. Kang, M.C. Stamm, A. Peng, K.J.R. Liu, Robust median filtering forensics using an autoregressive model, IEEE Trans. Inf. Forensics Secur. 8 (2013) 1456–1468.

[50] C. Quan, H. Lei, X. Sun, W. Bai, Multichannel convolutional neural network for biological relation extraction, BioMed Res. Int. 2016 (2016) 1–10.

[51] S. Ren, K. He, R. Girshick, J. Sun, Faster R-CNN: towards real-time object detection with region proposal networks, IEEE Trans. Pattern Anal. Mach. Intell. 39 (2015) 1137–1149.

[52] N. Srivastava, G. Hinton, A. Krizhevsky, I. Sutskever, R. Salakhutdinov, Dropout: a simple way to prevent neural networks from overfitting, J. Mach. Learn. Res. 15 (2014) 1929–1958.

[53] V. Nair, G.E. Hinton, Rectified linear units improve restricted boltzmann machines, in: International Conference on Machine Learning, 2010, pp. 807–814.

[54] P. Bas, T. Filler, T. Pevn, "Break our steganographic system": the ins and outs of organizing BOSS, in: International Conference on Information Hiding, 2011, pp. 59–70.

[55] J. Dong, W. Wang, T. Tan, CASIA image tampering detection evaluation database, in: IEEE China Summit & International Conference on Signal and Information Processing, 2013, pp. 422–426.

[56] J. Xu, Y. Ling, X. Zheng, Forensic detection of Gaussian low-pass filtering in digital images, in: 2015 8th International Congress on Image and Signal Processing (CISP), 2015, pp. 819–823.

About the authors

Dr. Bin Yang obtained his MS degree in Computer Science from South China University of Technology, China in 2007, and PhD degree in Computing Science and Technology from Hunan University, China, in 2014. From 2002 to 2010, he was a lecturer in South China Normal University. Since 2014, he has been an associate professor at the Jiangnan University, China. His research areas include information security, digital image forensic, deep learning and image processing. He has authored over 30 papers in relative areas.

Dr. Enguo Cao obtained his ME degree from Yanshan University in China in 2009, PhD degree from Kochi University of Technology in Japan in 2012. He was a research assistant in the Department of Intelligent Mechanical Systems Engineering, Kochi University of Technology, Kochi, Japan. He is currently an associate professor of Jiangnan University in China. His research interests include Cyber-Physical-Social Systems and deep learning.

Prof. Honglei Guo is currently a professor at the School of Design, Jiangnan University, Wuxi, China. He has study experience in northern Europe, Switzerland and other countries. His major research areas include image recognition, visual analysis and object detection. He has authored over 50 papers in relative areas.

CHAPTER THREE

Review on privacy-preserving data comparison protocols in cloud computing

Leqi Jiang[a,b], Zhihua Xia[a,b], and Xingming Sun[a,b]
[a]School of Computer and Software, Nanjing University of Information Science and Technology, Nanjing, China
[b]Jiangsu Engineering Center of Network Monitoring, Nanjing University of Information Science and Technology, Nanjing, China

Contents

Advances in Computers, Volume 120
ISSN 0065-2458
https://doi.org/10.1016/bs.adcom.2020.09.002

Abstract

Data comparison is a crucial step in many operations, including data mining, data classification, image feature extraction, and image retrieval. With the existing data encryption technology, encrypted data computation and encrypted data comparison can be
executed; however, there is no single data encryption method can meet the challenge
of tackle the two tasks at the same time. Therefore, in many data operations, two major
steps are involved: one is to encrypt the data with a computation supporting encryption
method; one is to construct a comparison protocol to conduct privacy-preserving data
comparison. While the first step has become a fast-developing and systematic research
spot, research in the second step area remains scattered. In this chapter, we focus on the
latter, by conducting an in-depth review of the development of the privacy-preserving
data comparison. We introduce and analyze several most commonly used privacy-
preserving data comparison protocols and their application scenarios; then we examine
their functionality, security, computation complexity and communication overhead. By
a systematic investigation of these protocols, we reveal and define the security challenge underneath each protocol. Our findings provide systematic and in-depth reference for improvement of the existing data comparison protocols in cloud environment.

Abbreviation

2PC	two-party computation
CS	cloud server
DO	data owner
DoG	difference of Gaussians
NHGD	negative hypergeometric distribution
OPE	order-preserving encryption
SHE	long short-term memory
SIFT	scale-invariant feature transform
SMC	secure multiparty computation
TP	trusty part
VIL	variable input-length
VOL	variable output-length

1. Introduction

Cloud servers, which have large storage space and strong computing
power, are widely used in big data era. Some well-known cloud servicers
include Alibaba Cloud, iCloud, IBM Cloud, etc. Individual users often
use cloud servers for storing their photos, diaries, or even some passwords;
Companies often use cloud servers to record and analyze their customers'
information to improve their service. The data uploaded to the cloud server

may contain a lot of sensitive information. In recent years, cloud server information leakage incidents occurred frequently.

In order to enhance cloud computing security, many encrypted data computation and encrypted data comparison methods had been established; however, up till now, there is no single data encryption method can meet the challenge of tackle the two tasks at the same time. Therefore, in many data operations, two major steps are involved: one is to encrypt the data with a computation supporting encryption method; one is to construct a comparison protocol to conduct privacy-preserving data comparison. In current landscape of encryption, the first step research area has become systematic from both theoretical and application perspectives; research on the latter remain scattered, lacking of theoretical development. In this chapter, we focus on investigating and analyzing the latter.

In the field of privacy-preserving data comparison protocols, Secure Multiparty Computation (SMC) approach is at the center. In SMC, mutually distrusting users can collaborate to compute a function of their private data without revealing any additional information. SMC approach has been utilized in data mining scenarios, where multiple users conduct the operation with no sensitive information leaked to each other [1]. For another instance, it is also utilized in digital images distribution scenario, where when a user wants to buy some image from the image owner, the cloud server with the owner's images sends the images to another cloud server, which embeds watermark on the images. The two servers system can protect the copyright and privacy of both parties [2]. Secure Multi-party data comparison (Yao's Millionaires' Problem) is a classic SMC problem [3]. It plays an important role in data mining [4,5], data classification [6,7], feature extraction [8,9], and image retrieval [10,11] etc. In this chapter, we conduct a systematic review on existing benchmark privacy-preserving data comparison protocols, which are built around this problem, in order to identify strength and weakness of each protocols, and further reveal and define the security challenge underneath each protocol.

The remainder of this chapter is organized as follows: An overview of privacy-preserving comparison is presented in Section 2. Section 3 introduces comparison protocols for curious data owners and review benchmark protocols under this category; Sections 4 and 5 introduce comparison protocols for curious cloud server with/without trusty part, and review benchmark protocols under this category. Section 6 compares the comparison protocols by category. In Section 7, we point out research direction in the field revealed by our study. Section 8 is the conclusion.

2. Issues in discussion

In order to enhance security of data on cloud servers, researchers have been seeking solutions from two perspectives: one is to encrypt the data with a computation supporting encryption method; one is to construct a protocol to conduct privacy-preserving data comparison. While some research on privacy-preserving data comparison protocols has been conducted [12–16], a systematic understanding of the comparative strength and weakness of these protocols become necessary for these protocols' application and further development.

The privacy-preserving data comparison protocols are built on the SMC. According to SMC, each participant owns a secret input, and all the participants want to jointly compute a function over their inputs while keeping those inputs private. SMC was introduced for the first time in a system of "Millionaires' Problem" raised by Yao [3]. The problem discusses two millionaires, Alice and Bob, who are interested in knowing who is richer without revealing their actual wealth. The solutions to Yao's millionaires' problem [17–19] were explored in cryptography. A number of privacy-preserving data comparison protocols have emerged, however, classification of these protocols has not been conducted in the exiting literature. In order to enable systematic view on the development of these protocols, we establish a classification system as in Fig. 1.

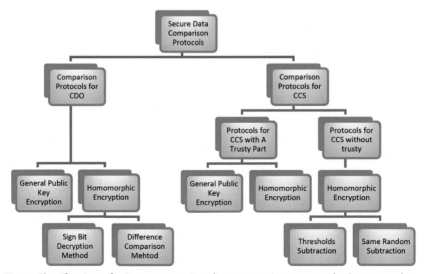

Fig. 1 Classification of privacy-preserving data comparison protocols. *Source: author.*

From the perspectives of major players in cloud security environment, i.e., users and servers, we classify these protocols into two categories: comparison protocols for Curious Data Owners (CDO) and comparison protocols for Curious Cloud Servers (CCS). Depending on whether the data owner's private key has been compromised, the protocols for CCS can be further divided into protocols with a trusty part and protocols without trusty part. Depending on the encryption scheme used, the solutions can be divided into two kinds: Data encrypted with general public key encryption [20–23] and data encrypted with homomorphic encryption [24–29]. When using general public key encryption, one has to decrypt the ciphertext got from the other. For security reasons, the ciphertext is always mixed in some interference items. That will increase the computing overhead and communication overhead. Homomorphic encryption can make the ciphertext calculation equivalent to the corresponding plaintext calculation, so the computing overhead and communication cost are less. We will review each component of the classification system in the following sections.

3. Comparison protocols for curious data owners (CDO)

The data owner A (DO A) owns plaintext x, and the other data owner B (DO B) owns plaintext y. They are all curious about the data the other one has. We define this data management scenario as comparison protocols for Curious Data Owners (CDO).

3.1 Framework of comparison protocols for CDO

The framework of comparison protocols for CDO can be constructed as the following: First, DO A encrypts his plaintext to a ciphertext and send it to DO B. Second, DO B encrypts his plaintext and use a specific comparison function to get the encrypted magnitude relation between their plaintexts. After DO B sends the result to DO A, DO A decrypt it and obtain the magnitude relation. The general model of the comparison protocols for curious data owners is illustrated in Fig. 2.

- $E(\cdot)$ for Encryption.
- $D(\cdot)$ for Decryption.
- $F(\cdot)$ for one specific sorting scheme.
- c for the encrypted magnitude relation.
- m for the magnitude relation.

Depending on the encryption algorithm, the comparison protocols for CDO can be divided into two categories: protocols with general public key encryption and protocols with homomorphic encryption. In the early

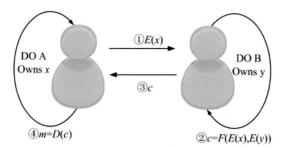

Fig. 2 The general model of the comparison protocols for curious data owners. *Source: author.*

Table 1 Yao's comparison protocol.
Algorithm 1: Yao's comparison protocol

Input: DO A inputs x and DO B inputs y.
Output: The magnitude relation between x and y.
At DO A:
Generates public/private key pair (pk, sk);
Shares pk to DO B;
At DO B:
Picks a random L-bit nonnegative integer a, and encrypts it to get $k = E_{pk}(a)$;
Sends DO A the number $k - y + 1$;
At DO A:
Computes the values of $I_u = D_{sk}(k - y + u)$, for $u = 1, 2, ..., N$;
Keeps generating a random prime p of $L/2$ bits, and computing the values $z_u = I_u(mod\ p)$ for all u until all I_u differ by at least 2 in the $mod\ p$ sense;
Sends the prime p and the following N numbers to DO B: $z_1, z_2, ..., z_x$ followed by $z_{x+1} + 1, ..., z_N + 1$; the above numbers should be interpreted in the $mod\ p$ sense.
At DO B:
Looks at the y-th number (not counting p) sent from DO A, and decides that $x \geq y$ if it is equal to $a\ mod\ p$, and $x < y$ otherwise;
Share the result to DO A.

Source: author.

years when Yao's Millionaires' Problem was first proposed, there were many protocols designed with general public key encryption. Afterward, the homomorphic encryption algorithm became more trendy.

3.2 Protocols with general public key encryption

As an initiator, Yao proposed a simple scheme to resolve the Millionaires' Problem in Ref. [3]. This protocol proceeds are given in Table 1, and the processes is shown in Fig. 3. When DO A and B compare the magnitude

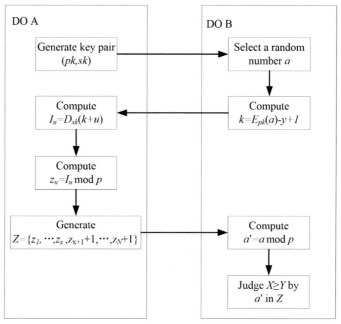

Fig. 3 The processes of Yao's comparison protocol. *Source: author.*

of X and Y they have, DO B generate a random number a and encrypt it. After that, the ciphertext of a will substract Y. DO A will add 1 to X to the result and decrypt them. If a is shown in the final results, they can get $X \geq Y$, or $X < Y$. This scheme conducts comparison between two owners with small-sized data. It cannot process large volume of data, while the effectiveness and security of the protocol was Yao's focus.

This scheme achieved security by the following mechanism: The random prime p ensures that there is no identical elements of the vector passed from DO A to DO B. And no plaintext associated directly with x is transmitted through the scheme. Although DO A gets the number $k - y + 1$, he has no information about k, so he could not derive the DO B's plaintext y. In the case where DO A/B cannot derive the other party's plaintext from the transmitted data, there is no special requirement on the encryption algorithm. Another factor that determines the scheme's security is the size of N. If the N is large enough, the scheme will be safe.

However, while the security level rising of the scheme, the communication overhead is also increasing. Fig. 3 shows that in this scheme, three data transmissions between the two data owners are required before the magnitude relation is obtained. The users need to undertake a lot of calculation tasks. The detailed quantitative analysis will be given in Section 3.4.

3.3 Protocols with homomorphic encryption

Homomorphic encryption is a form of encryption that allows computation on ciphertexts. And the decrypted result matches the result of the operations as if they had been performed on the plaintext. To reduce the communication overhead and computation overhead, homomorphic encryption is chosen to encrypt the data. Homomorphic encryption can support addition, subtraction and multiplication in encrypted domain as formulas (1)–(3).

$$E(x + y) = E(x) \oplus E(y) \tag{1}$$
$$E(x \times y) = E(x) \otimes E(y) \tag{2}$$
$$E(x - y) = E(x) \oplus E(-1) \otimes E(y) \tag{3}$$

3.3.1 Sign bit decryption method

Lin [24] proposed a two-round solution to Yao's Millionaires' Problem based on homomorphic encryption. The plaintexts kept by DOs are expressed as k-bit bitstreams. In Lin's protocol, two special encodings, 0-encoding and 1-encoding are used. x is greater than y if x's 1-encoding has a common element with y's 0-encoding. Let $s = s_n s_{n-1} \ldots s_1 \in \{0,1\}^n$ be a binary string of length n. The 0-encoding of s is the set S_s^0 of binary strings such that

$$S_s^0 = \left\{ s_n s_{n-1} \ldots s_{i+1} 1 \,|\, s_i = 0, 1 \leq i \leq n \right\}$$

The 1-encoding of s is the set S_s^1 of binary strings such that

$$S_s^1 = \left\{ s_n s_{n-1} \ldots s_{i+1} s_i \,|\, s_i = 1, 1 \leq i \leq n \right\}$$

Give an example: Let $x = 6 = 110_2$ and $y = 2 = 010_2$ we have $S_x^1 = \{1, 11\}$, $S_x^0 = \{111\}$ and $S_y^0 = \{1, 011\}$, $S_y^1 = \{01\}$. So we can get $x > y$ because $S_x^1 \cap S_y^0 \neq \varnothing$, else $x \leq y$. Since the two data owners use the same encoding rules, in order to prevent the other one from getting the codes and deriving the original data, the intersection of their encodings are obtained in the encrypted domain. The protocol proceeds are described in Table 2. The processes in Lin's protocol is shown in Fig. 4.

Compared with Yao's protocol, Lin's scheme has one less data transmission between two owners. This protocol supports data comparison with a larger range. However, the two special encodings methods greatly increase the amount of plaintexts which need to be encrypted. In addition, the number of ciphertexts is also greatly increased due to the encryption methods.

Table 2 Lin's comparison protocol.
Algorithm 2: Lin's comparison protocol

At DO A:

Generates a homomorphic encryption key pair (pk, sk);

Get the 1-encoding of X as $S_x^1 = \{x_k x_{k-1} \ldots x_1\}$;

For each element in S_x^1, compute $T[x_j, j] = E_{pk}(1)$ and $T[\overline{x}_j, j] = E_{pk}(r_j)$ for some random r_j, $1 \leq j \leq k$; where $x_j = 1$ or 0, and $\overline{x}_j = 1 - x_j$;

Builds the n $2 \times k$-tables $T[i, j]$;

Sends the tables and pk to DO B;

At DO B:

Computes $c_t = T[t_k, k] \odot T[t_{k-1}, k-1] \cdots \odot T[t_i, i]$ for each $t = t_k t_{k-1} \cdots t_i \in S_y^0$, where $\odot$ denotes inclusive-or;

Permutes c_t's randomly as $\{c_l\}$; there are $|S_x^1| \times |S_y^0|$ elements in $\{c_k\}$ because all elements in S_x^1 should be compared with all elements in S_y^0.

Sends $\{c_k\}$ to DO A;

At DO A:

Decrypts $D(c_i) = m_i$, and determines $x > y$ if and only if some $m_i = 1$;

Shares the result to DO B.

Source: author.

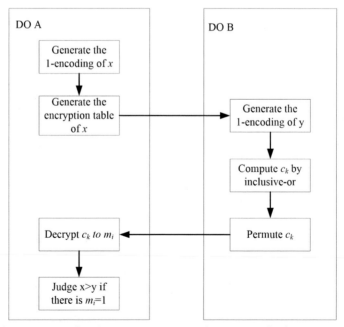

Fig. 4 The processes of Lin's comparison protocol. *Source: author.*

Hu [25] proposed a privacy-preserving comparison protocol applied in image Scale-Invariant Feature Transform (SIFT) extraction. Although the participants in this protocol are two cloud servers, it was classified as a comparison protocol for CDO rather than for CCS. Hu outsourced the processing of image SIFT feature extraction to the cloud. He split a value into the difference of two values to preserve data safe, and obtained two matrices of the same size as the original image. He sent the two matrices to two different cloud servers separately. Somewhat Homomorphic Encryption (SHE) is used in his scheme. And this scheme calculates the magnitude relationship of two values by bit in binary. Although, the data kept by cloud servers are ciphertexts of data owner's image, they are plaintext to the two cloud servers when locating the keypoints. Therefore, despite the name of participants is cloud server, the actual role of them is data owner.

In Ref. [25], data is also expressed as n-bit bitstreams by binarization, $x = x_n x_{n-1} \ldots x_1$ and $y = y_n y_{n-1} \ldots y_1$. Then for $i = k, k-1, \ldots, 1$, he computes a_i and b_i as follows.

- If $i = k$, then $a_i = x_i \times (1 - y_i)$, $b_i = y_i \times (1 - x_i)$;
- If $i < k$, then

$$a_i = (1 - b_{i+1}) \times (a_{i+1} + (1 - a_{i+1}) \times x_i \times (1 - y_i))$$
$$b_i = (1 - a_{i+1}) \times (b_{i+1} + (1 - b_{i+1}) \times y_i \times (1 - x_i))$$

If the final result $a_1 = 1$, we can get x is greater than y. Those can be easily achieved by SHE. The details is shown in Table 3. The processes of this protocol is shown in Fig. 5.

Similar to Lin's protocol, two data transmissions between the two data owners are required before the magnitude relation is obtained in Hu's protocol. Because the data are expressed by binarization rather than the special encodings in Ref. [24], the data transmitted are much less. However, large volume of calculation is involved in the solution process of a_1. As the data owners are cloud servers in Ref. [25], this protocol is effective by making full use of the cloud servers.

3.3.2 Difference comparison method

Compared with Yao's protocol, Lin's protocol has reduced communication overhead. However, big load of work is still demanded from data owners in this protocol. As the load of work of both comparison protocols is depending on the data under comparison, it is uncontrollable. Kumar [26] proposed an efficient solution to resolve this problem. His protocol determines the magnitude relation between two plaintexts by judging whether the difference between the two was greater than zeros. The encrypted difference can be

Table 3 Hu's comparison protocol.
Algorithm 3: Hu's comparison protocol

At DO A:

Generates a homomorphic encryption key pair (pk, sk);

Gets the binary of x as $x = x_k x_{k-1} \ldots x_1$;

Encrypts each bit of x separately, $c_i = E_{pk}(x_i)$, $1 \leq i \leq k$;

Sends the ciphertexts and pk to DO B;

At DO B:

Encrypts 1 and -1 as $T = E_{pk}(1)$ and $T' = E_{pk}(-1)$;

Gest the binary of y as $y = y_k y_{k-1} \ldots y_1$;

Encrypts each bit of x separately, $w_i = E_{pk}(y_i)$, $1 \leq i \leq k$;

for $i = k, k-1, \ldots, 1$ do

if $i = k$ then

$a_i = c_i \otimes (T \oplus T' \otimes w_i)$;

$b_i = w_i \otimes (T \oplus T' \otimes c_i)$;

else

$a_i = (T \oplus T' \otimes b_{i+1}) \otimes (a_{i+1} \oplus (T \oplus T' \otimes a_{i+1}) \otimes c_i \otimes (T \oplus T' \otimes w_i))$;

$b_i = (T \oplus T' \otimes a_{i+1}) \otimes (b_{i+1} \oplus (T \oplus T' \otimes b_{i+1}) \otimes w_i \otimes (T \oplus T' \otimes c_i))$;

end if

end for

Send the final a_1 to DO A.

At DO A:

Decrypts a_1 to get the magnitude relation.

Source: author.

directly obtained using homomorphic. To ensure that data owners cannot derive another number by the difference and one of the numbers. The difference is multiplied by a random number before decryption. The protocol is described in Table 4. The processes of Kumar's comparison protocol is shown in Fig. 6.

Compared with Yao's scheme, this scheme is much simpler, and the cost of communication and computation is lower. Although it requires the same communications times as Lin's comparison protocol, the data amount transmitted in the protocol is greatly reduced. This protocol is secure, as homomorphic encryption is nondeterministic encryption and it can protect against ciphertext-only attacks. Despite $E_{pk}(x)$ is transmitted to DO B, he could not derive x from $E_{pk}(x)$. By homomorphic operations, DO A can only get the plaintext $r \times (x - y)$ which will not leak y or the difference between x and y.

3.4 Analysis of comparison protocols for CDO

In this section, we conduct a quantitative analysis on the security and performance of the comparison protocols for CDO we have reviewed.

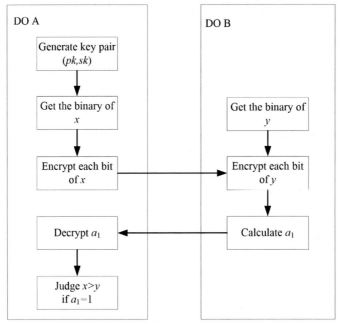

Fig. 5 The processes of Hu's comparison protocol. *Source: author.*

Table 4 Kumar's comparison protocol.
Algorithm 4: Kumar's comparison protocol

At DO A:

Generates a homomorphic encryption key pair (pk, sk).

Encrypts x to $E_{pk}(x)$;

Shares pk and $E_{pk}(x)$ to DO B;

At DO B:

Chooses a random number r and encrypts r, y to $E_{pk}(r)$, $E_{pk}(y)$ with DO A's public key;

Computes $V = E_{pk}(r) \otimes (E_{pk}(x) - E_{pk}(y)) = E_{pk}(r \times (x - y))$;

Sends V to DO A;

At DO A:

Decrypts V to $m = D_{sk}(V)$;

Sends the most significant bit of m to DO B;

At DO B:

Takes XOR of the obtained bit from DO A and the MSB of r to obtain the output: if $x > y$;

Share the result to DO A.

Source: author.

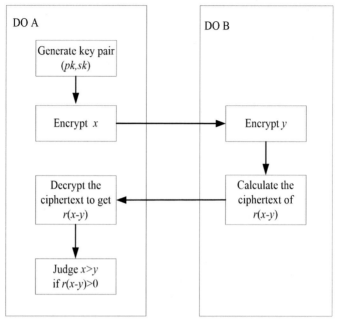

Fig. 6 The processes of Kumar's comparison protocol. *Source: author.*

3.4.1 The security of the protocols

The participants in comparison protocols for CDO are considered as honest but curious. They follow the protocols honestly, but they try to acquire extra information about the other party's plaintext. The only requirement for the encryption is to be able to resist ciphertext-only attacks. Therefore, the mayor issues under our investigation is how much information about the data owner's data will be obtained by other participants, and what is the probability that the data owner's data being acquired.

Assume there are two numbers x, y (DO A owns x and DO B owns y) to be compared with each other, $x, y \in [1, N]$. Because the data owner will not try to decrypt the ciphertexts without private key, only the leakage of plaintext information and the probability of the participants deriving the plaintext are considered in security analysis. The analysis result is shown in Table 5.

The protocols in [3,24,25] guarantee no plaintext information being leaked to others. Although "$r \bullet (x - y)$" is leaked to DO A, r is a random number selected by DO B, DO A cannot derive y. x, y are integers in the range of 1 to N, so "$1/N$" means DO A/B could not derive the plaintext

Table 5 Security analysis of protocols for CDO.

Protocols	Data leaked to DO A	Data leaked to DO B	Probability of DO A deriving plaintext	Probability of DO B deriving plaintext
Yao [3]	None	None	$1/N$	$1/N$
Lin [24]	None	None	$1/N$	$1/N$
Hu [25]	None	None	/	/
Kumar [26]	$r \bullet (x - y)$	None	/	/

Source: author.

Table 6 Performance of protocols for CDO.

Protocols	Computation complexity	Communication overhead	
		Data amount	Transfer times
Yao [3]	$O(n \bullet N)$	$(1 + N) \bullet C_n^2$	$3 \bullet C_n^2$
Lin [24]	$O(n \bullet k)$	$4 \bullet k \bullet C_n^2$	$2 \bullet C_n^2$
Hu [25]	$O(n)$	$2 \bullet C_n^2$	$2 \bullet C_n^2$
Kumar [26]	$O(n)$	$2 \bullet C_n^2$	$2 \bullet C_n^2$

Source: author.

y/x from known information. What is more, "/" means DO A/B could not get the range of DO B/A's data in Refs. [25,26]. These comparison protocols for CDO are proved to be secure.

3.4.2 The performance of the protocols

The two indicators for the performance of the comparison protocols are computational complexity and communication overhead. Communication overhead consists data amount (transmission of public key is not considered) transmitted between participants and transfer times. Both of them count from the beginning of the protocol until one participant obtains the magnitude relation. We assume there are n numbers to be compared with each other. The performance of protocols for curious data owners is shown in Table 6.

The computation complexity of protocols with general public key encryption is higher than protocols with homomorphic encryption. And more data is transmitted in the protocols with general public key encryption.

4. Comparison protocols for curious cloud servers (CCS) with a trusty part

Cloud server has been widely used due to its strong computation ability and huge storage capacity. Data owner use cloud server to store his data and process them. However, for the security reasons, the data will be uploaded after encrypting [30]. Kaghazgaran named Comparison protocols for CCS as comparison over encrypted data [31]. Kaghazgaran's model is the first problem model different from Yao's problem. It provides solutions to the problem that "if a person owns a lot of data, how to design a protocol that allow others to help him on comparison but without leaking his information" [31].

4.1 General model of comparison protocols for CCS with a trusty part

Within comparison protocols for CCS, one type involves a third party in the process. Data Owner (DO) will not take part in the comparison except providing the data x_i to be processed after encryption. Cloud Server (CS) is used to store DO's data. A Trusty Part (TP) is used to help decrypt the ciphertexts into comparable plaintexts. It holds the data owner's private key sk. Both CS and TP are curious about DO's data.

The framework of comparison protocols for CCS is described as follows: First, DO sends his ciphertexts to CS and shares his private key to TP. Second, CS does some transformations on the ciphertexts to make sure TP will not get DO's original data. Third, CS sends the transformed ciphertexts to TP. Fourth, TP decrypts the ciphertexts with DO's private key and sorts the decrypted data. Fifth, TP shares the sorting results to CS. The general model of those comparison protocols is illustrated in Fig. 7.

- (pk, sk) for public/private key pairs generated by data owner.
- $E_{pk}(\bullet)$ for encryption with pk.
- $D_{sk}(\bullet)$ for decryption with sk.
- c_i is ciphertext of $E_{pk}(x_i)$. It can be sorted by decrypting with sk, but will not leak x_i.

Comparison protocols for CCS with a trusty part also can be divided into two categories: protocols with general public key encryption and protocols with homomorphic encryption.

4.2 Protocols with general public key encryption

Kaghazgaran's first protocol used public key encryption [31]. The DO uploads the ciphertexts of his data to CS. CS selects a random integer l

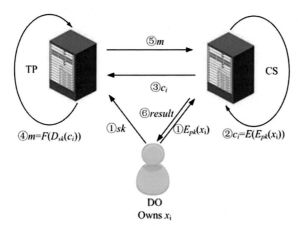

Fig. 7 The general mode of comparison protocols for server to server. *Source: author.*

Table 7 Kaghazgaran's protocols with general public key encryption.
Algorithm 5: Kaghazgaran's first comparison protocol

At DO:
Generates a public/private key pair (pk, sk);
Shares the private key sk to TP;
Uploads $E_{pk}(x_i)(i = 1, 2, \cdots, n)$, the ciphertexts of his data x_i's, to CS;
At CS:
Picks a random integer $l(1 < l < L)$ and computes the values of: $c_i = E_{pk}(x_i) - l$;
Sends c_i's to TP;
At TP:
Decrypts c_i's to get $m_{i,j} = D_{sk}(c_i + j)(j = 1, 2, \cdots, L)$;
Generates random r and computes $M_{i, j} = m_{i,j} + r$;
Sends the n arrays in size of L to CS;
At CS:
Looks at l-th number of each array sent by CS A, and get the sort of the data it kept.

Source: author.

and calculates the different of ciphertexts and the integer, then the results will be sent to TP. TP recovers the plaintexts. TP adds 1 to N to the result in order, without knowing the integer, and decrypt them. CS gets the magnitude relation from the l-th result. The protocol is described in Table 7. The processes is shown in Fig. 8.

This scheme is similar to Yao's scheme. Although TP can get $m_{i,l} = x_i, l$ is not known by TP, it cannot derive which ones are the data owner's original data. The operation $M_{i,j} = m_{i,j} + r$ makes sure that CS can sort those data without knowing the exact values of DO's original data. This protocol takes

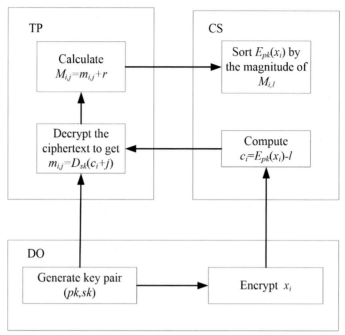

Fig. 8 The processes of Kaghazgaran's protocols with general public key encryption. *Source: author.*

full advantage of cloud servers, and makes it possible to sorting over encrypted data without data owner involved. However, it still comes with big computation overhead and communication overhead.

4.3 Protocols with homomorphic encryption

Although Kaghazgaran's first comparison protocol [31] enabled comparison over encrypted data without leaking DO's original data, the communication overhead between CS and TP is high. In order to improve the efficiency, he proposed another protocol based on homomorphic encryption. In this protocol, CS uses homomorphic cryptosystem to add an identical random number before the ciphertexts being sent to the TP. In this way, the magnitude relationship of the result does not change after decryption, and the original plaintext is protected. The details and process of protocol are shown in Table 8 and Fig. 9.

Although this protocol takes more steps than Kaghazgaran's protocols with general public key encryption, the computing burdens upon the servers are less. This protocol is secure as data are protected by only adding a random

Table 8 Kaghazgaran's protocols with homomorphic encryption.
Algorithm 6: Kaghazgaran's second protocol

At DO:

Generates a homomorphic encryption public/private key pair (pk, sk);

Shares the private key sk to TP;

Uploads $E_{pk}(x_i)(i=1,2,\cdots,n)$, the ciphertexts of his data x_i's, to CS;

At CS:

Picks a random integer R and encrypts it to $E_{pk}(R)$;

Computes $c_i = E_{pk}(x_i) \oplus E_{pk}(R)$ $(i=1,2,\cdots,n)$;

Sends c_i's to TP;

At TP:

Decrypts them to gets $x_i + R$ and sorts them;

Shares the sorting table to CS.

Source: author.

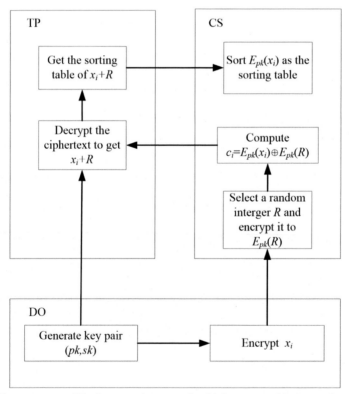

Fig. 9 The processes of Kaghazgaran's protocols with homomorphic encryption. *Source: author.*

number instead of mixed with a lot of random numbers. Although TP owns DO's private key, it can only get c_i which are the ciphertexts corresponding to $x_i + R$. It cannot get the exact value of x_i without knowing R. What is more, TP only shares the sorting table to CS, which does not include any information about DO's original data's exact values.

4.4 Analysis of comparison protocols for CCS with a trusty part

In this section, we conduct a quantitative analysis on the security and performance of the comparison protocols for CCS with a trusty part we have reviewed.

4.4.1 The security of protocols

Assume there are n numbers in the range of 1 to N to be sorted. The security analysis result is shown in Table 9.

In Table 9, x_i mixed with r_j denotes TP could get n vectors, each vector includes one plaintext of x_i and $L - 1$ different random numbers r_j. R, r are two random numbers.

TP know what x_i's are with $1/L$ probability in the first protocol, so the appropriate L needs to be selected to balance the relationship between security and computation complexity in the first scheme. In the second scheme, it is difficult for TP to guess the real values of x_i without any information about the random integer R.

4.4.2 The performance of protocols

As DO is the provider of encrypted data and encryption key pair in these protocols, only the overheads of communication between TP and CS are investigated in Table 10. As CS is the result receiver, the communication overhead counts from the beginning of the protocol until CS obtains the magnitude relation.

Table 9 Security analysis of protocols for CCS with a trusty part.

Protocols	Data leaked to TP	Data leaked to CS	Probability of TP deriving plaintext	Probability of CS deriving plaintext
Kaghazgaran [31].1	x_i mixed with r_j	$r + x_i$	$1/L$	/
Kaghazgaran [31].2	$R + x_i$	None	/	/

Source: author.

Table 10 Performance of protocols for CCS with a trusty part.

Protocols	Computation complexity	Communication overhead	
		Data amount	Transfer times
Kaghazgaran [31].1	$O(n \cdot L)$	$n \cdot L + L$	2
Kaghazgaran [31].2	$O(n)$	$2 \cdot n$	2

Source: author.

For security, L is a number much bigger than 2. So the protocol with homomorphic encryption proposed by Kaghazgaran has lower computation complexity and communication overhead than the protocol with general homomorphic encryption.

5. Comparison protocols for CCS without trusty part

Cloud servers are all considered to be honest but curious. Treating them as a trusty part and giving them the private key will cause security problems. Therefore, comparison protocols for curious cloud servers with a trusty part are not practical. Many comparison protocols for curious cloud server without trusty part were proposed in recent years [32–34].

5.1 General model of comparison protocols for CCS without trusty part

Similar to those protocols with trusty part, the protocols without trusty part consist of three participants—data owner (DO), cloud server A (CS A) and cloud server B (CS B). The difference is that CS A is no longer a trusty part and it owns an incomplete decryption algorithm $D'(\cdot)$ and a specific sorting algorithm $F(\cdot)$ rather than DO's private key sk. After incomplete decryption, the ciphertext is transformed to data, which is sorted with $(\cdot)$, but does not reflect the original plaintext.

The framework of comparison protocols for CCS without trusty part is described as follows: First, DO sends his ciphertexts to CS B and shares an incomplete decryption algorithm and a specific sorting algorithm to CS A. These two algorithms are unique algorithms related to the encryption DO used and DO's private key. Second, CS B does some transformations on the ciphertexts to make sure CS A will not get DO's original data. Third, CS B sends the transformed ciphertexts to CS A. Fourth, CS A decrypts the ciphertexts with the incomplete decryption algorithm and sorts them with the specific sorting algorithm. Fifth, CS A shares the comparison results to CS B. The general model of those comparison protocols is illustrated in Fig. 10. For some incomplete decryption algorithms, $c_i = E(E_{pk}(x_i))$ can be omitted. So there are many protocols only consist of two roles: DO and CS.

- (pk, sk) for public/private key pairs generated by data owner.
- $E_{pk}(\bullet)$ for encryption with pk.
- $D_{sk}(\bullet)$ for decryption with sk.
- c_i is ciphertext of $E_{pk}(x_i)$. It can be sorted by decrypting with sk, but will not leak x_i.
- $F(\bullet)$ for specific sorting algorithm.
- $D'(\bullet)$ for incomplete decryption algorithm.

Due to the limitations of $F(\bullet)$, the comparison protocols without trusty part has high requirements on the encryption algorithm. Only homomorphic encryption algorithms can meet the requirements.

5.2 Protocols with homomorphic encryption

The random numbers are introduced during encryption to enable the magnitude relation of those ciphertexts not to be directly known. Those random numbers may make $E(x) > E(y)$ even if $x < y$. Therefore, comparison protocols for CCS without trusty part usually eliminate the influence of random numbers to complete the comparison. Hsu [27] proposed to judge the magnitude relation by comparing the distance between different ciphertexts and the thresholds. Based on Hsu's work, Bellafqira [28] proposed a protocol

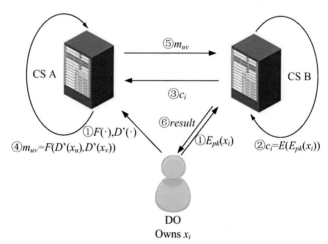

Fig. 10 The general model of comparison protocols for CCS without trusty part. *Source: author.*

with only one threshold, where the magnitude relation can be obtained by knowing the distance between the threshold and the ciphertext. Jiang [29] proposed that the numbers can be encrypted once more by the other ciphertext's random numbers, so the magnitude relation can be directly obtained by comparing the magnitude of the values of the ciphertexts.

5.2.1 Thresholds subtraction method

Hsu [27] was the first to propose outsourcing SIFT feature extraction of image to the cloud server. He choose Paillier cryptosystem to protect his original image. Since each encryption of Paillier uses a different random number, the plaintext and ciphertext are not mapped one by one. Hsu analyzed the calculations on ciphertext and found that the random numbers $r_{i,j}$ used in the initial ciphertext have the following relationship with the random numbers R_ρ of the intermediate results: $R_\rho = \prod_{u,v} r_{i-u,j-v}^{G_{Diff}(u,v,\rho)} (G_{Diff}(u,v,\rho)$

is a public parameter in SIFT. He proposed a method for eliminating the influence of random numbers, so that the result could be sorted as the plaintext.

In his scheme, a series of threshold T_k are encrypted with the same random numbers as the ciphertexts, which are to be compared. In this way, it can map the magnitude relationship between ciphertexts into the distance of

Table 11 Hsu's comparison protocol.
Algorithm 7: Hsu's comparison protocol

At DO:
Generates a homomorphic encryption public/private key pair (pk, sk);
Encrypts his image with random numbers $r_{i,j}$;
Computes $R_\rho = \prod_{u,v} r_{i-u,j-v}^{G_{Diff}(u,v,\rho)}$;
Generates a series of threshold T_k;
Encrypts each of the thresholds with different R_ρ;
Sends the ciphertexts $E(x_{i,j}, r_{i,j})$ and $E(T_k, R_\rho)$ to CS, where $E(x, r)$ denotes the plaintext x is encrypted with random number r;
At CS:
Computes $E(x_{i,j}, r_{i,j})$ to $E(x_{i,j}', R_\rho)$;
Computes $a_l = \underset{\forall k}{argmin}\left(E\left(x_{i,j}', R_\rho\right) - E\left(T_k, R_\rho\right)\right)$ to get the distance of $x_{i,j}'$ to T_k;
Sorts the distance to get the magnitude relation of $x_{i,j}'$.

Source: author.

plaintext to the thresholds. The details and the process of the protocol are shown in Table 11 and Fig. 11.

This is a secure protocol to outsource all processes of SIFT feature extraction to the cloud server. DO only needs to communicate with the server once. However, the DO is required to take some extra work, such as calculating R_ρ and encrypting the thresholds T_k. Therefore, this protocol requires DO to have a certain degree of understanding of data processing. This makes it not easy to be applied directly to other problems. In addition, the DO's workload increases, as the number of thresholds increases or the size of the picture increases.

In order to reduce the computing burden of cloud server users, some researchers proposed to encrypt the thresholds with the same random numbers as the image. After the same data processing, the influence of the random numbers on the magnitude relation can be eliminated. Bellafqira [28] proposed a solution using one threshold for the deficiencies in Ref. [27]. He introduced two methods: one is to compare data encrypted with same public key, another is to compare data encrypted with different public keys. The details and process of his protocol with same public key are shown in Table 12 and Fig. 12.

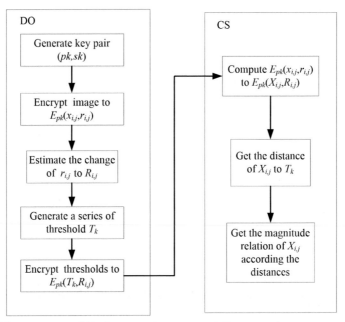

Fig. 11 The processes of Hsu's comparison protocol. *Source: author.*

Table 12 Bellafqira's comparison protocol with same public key.
Algorithm 8: Bellafqira's comparison protocol with same public key

At DO:

Generates a homomorphic encryption public/private key pair (pk, sk);

Encrypts x_j with public key to get $E_{pk}(x_j, r_i)$, where $E_{pk}(x, r)$ denotes the plaintext x is encrypted with random number r by the public key pk;

Generates a threshold T;

Encrypts the threshold T with the same random numbers to get $E_{pk}(T, r_i)$;

Sends $E_{pk}(T, r_i)$ to CS.

At CS:

Computes $D_{aT} = E_{pk}(x_a, r_i) - E_{pk}(T, r_i)$ and $D_{bT} = E_{pk}(x_b, r_i) - E_{pk}(T, r_i)$;

Gets $x_a > x_b$ if $D_{aT} > D_{bT}$.

Source: author.

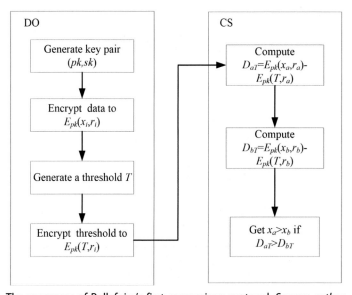

Fig. 12 The processes of Bellafqira's first comparison protocol. *Source: author.*

This protocol is simpler than Hsu's protocol. The distance can be get if the threshold encrypted with the same public key and random number with DO's plaintexts encrypted. Although $x_a - T \neq D_{aT}$, $D_{aT} - D_{bT}$ has the same sign bit as $x_a - x_b$. CS can only get D_{aT}, which will not leak any information about x_a, even if T is obtained by CS.

Based on this method, Bellafqira proposed another comparison protocol, which can obtain the magnitude relation between two ciphertexts encrypted with different public keys. The details and process of the protocol with

Table 13 Bellafqira's comparison protocol with different public keys.
Algorithm 9: Bellafqira's comparison protocol with different public keys

---

At DO:

Generates homomorphic encryption public/private key pairs (pk_j, sk_j);

Encrypts x_a with public key pk_A to get $E_{pk_A}(x_a, r_i)$ and x_b with public key pk_B to get $E_{pk_B}(x_b, r_i)$;

Generates a threshold T;

Encrypts the threshold T with the different public keys to get $E_{pk_A}(T, r_i)$ and $E_{pk_B}(T, r_i)$;

Sends $E_{pk_A}(x_a, r_i)$, $E_{pk_B}(x_b, r_i)$, $E_{pk_A}(T, r_i)$, $E_{pk_B}(T, r_i)$ to CS.

At CS:

Computes $D_{aT} = E_{pk_A}(x_a, r_i) - E_{pk_A}(T, r_i)$ and $D_{bT} = E_{pk_B}(x_b, r_i) - E_{pk_B}(T, r_i)$;

Gets $x_a > x_b$ if $D_{aT} > D_{bT}$.

Source: author.

different public keys are shown in Table 13 and Fig. 13. Two different ciphertexts subtract to the same threshold encrypted with the corresponding random number, and their magnitude relationship can be obtained by comparing the differences.

Although these two protocols can reduce computation overhead of DO, they are not of high security level: while the influence of random numbers is eliminated, the matrix formed by the difference between all the plaintext and the same threshold can also reflect the information of the original image. In addition, these two protocols did not take into consideration the situation where the data might be processed in the ciphertexts domain.

5.2.2 Same random subtraction method

Instead of using the threshold as a medium for random number subtraction, Jiang [29] proposed a method for directly encrypting the plaintext using the same random number. His method encrypts the data with level homomorphic encryption. It encrypts plaintext m to ciphertext c using the following formula:

$$c = \left[\lfloor q/t \rfloor [m]_t + e_i + h s_i \right]_q \tag{4}$$

where q, t, h are public parameters, e_i and s_i are random numbers. $\lfloor x \rfloor$ denotes the max integer, which is less than x, $[x]_t$ equals $x \bmod t$.

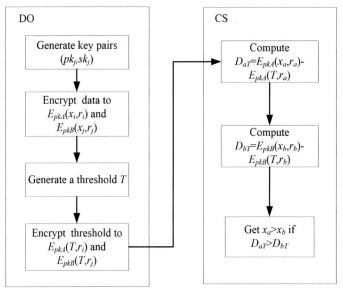

Fig. 13 The processes of Bellafqira's second comparison protocol. *Source: author.*

Table 14 Jiang's comparison protocol.
Algorithm 10: Jiang's comparison protocol

At DO:
Encrypts m_1 with random numbers e_1, s_1 to get $E(m_1, e_1, s_1)$;
Encrypts m_2 to get $E(m_2, e_2, s_2)$;
Encrypts m_2 with random numbers e_1, s_1 to get $E(m_2, e_1, s_1)$;
Sends $E(m_1, e_1, s_1)$, $E(m_2, e_2, s_2)$, $E(m_2, e_1, s_1)$ to CS;
At CS:
Computes $D_{m_{12}} = E(m_1, e_1, s_1) - E(m_2, e_1, s_1)$;
Gets $m_1 > m_2$ if $D_{m_{12}} > 0$.

Source: author.

The magnitude of two plaintexts can be obtained by ciphertexts subtraction, if their ciphertexts are encrypted with same random numbers. The details and process of Jiang's protocols are shown in Table 14 and Fig. 14.

Jiang's protocol is the simplest in form, but it is hard to be in to practice. In order to support the comparison of intermediate results, the DO must consider all combinations of original data and encrypt them using different random numbers. Compared with the scheme in Ref. [27], this solution greatly increases the workload of DO.

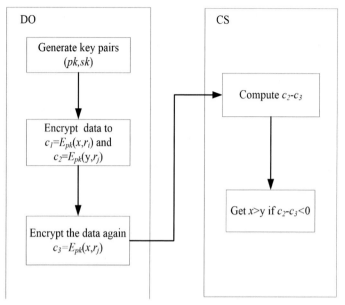

Fig. 14 The processes of Jiang's comparison protocol. *Source: author.*

5.3 Analysis of comparison protocols for CCS without trusty part

In this section, we conduct a quantitative analysis on the security and performance of the comparison protocols for CCS with a trusty part we have reviewed.

Assume there are n numbers in the range of 0 to $N-1$ to be sorted. L denotes the numbers of thresholds in [27]. The result of security analysis and performance analysis is shown in Table 15.[a]

Without knowing the value of threshold T, CS cannot derive all the plaintexts with existing information in the three protocols. However, the protocols proposed in them is to deal with image data. In Ref. [28] x_i is hardly to be derived when i equals to a certain value. But it is easy to retrieve the original image when putting all $x_i - T$ in a matrix. The protocol in [27] is more secure with multi-threshold. But DO has to undertake more workload. Although the protocol in [29] is secure, it greatly increases DO's

---

[a] Bellafqira's two protocols [28] are of the same level of security and effectiveness. As in Algorithm 9, DO encrypts its data with different public keys, which increases its own calculations and difficulty of keys management, this algorithm is not practical. Therefore, we only include Algorithm 8 in Bellafqira [28] in our investigation.

Table 15 Analysis of protocols for CCS without trusty part.

Protocols	Data leaked to CS	Probability of CS deriving plaintexts	Computation complexity	Communication overhead Data amount
Hsu [27]	$x_i - T_j$	/	$O(n \cdot L)$	$< n \cdot N + n$
Bellafqira [28]	$x_i - T$	/	$O(n)$	$2 \cdot n$

Source: author.

workload. L in [27] is the number of thresholds, which can be controlled by DO, it cannot increase whether or not n increases. In contrast, the number of data n cannot be controlled. When facing a large amount of data, DO in Ref. [29] undertake much more work than DO in Ref. [27].

5.4 Order preserving encryption

Order Preserving Encryption (OPE) is an encryption scheme, where the sort order of ciphertexts match the sort order of the corresponding plaintexts [35,36]. With OPE, the magnitude relation of ciphertexts can be easily obtained without additional comparison protocols. Therefore, OPE is a good tool for comparison protocol for CCS without trusty part.

Qin [37] proposed a secure SIFT extraction method based on OPE. In his method, the images are split into two ciphertext matrices. Those two matrices are sent to two different cloud servers, and the cloud servers can collaborate to complete extreme points detection.

Although OPE performs well in ciphertexts magnitude relation acquisition, it cannot easily process data in the ciphertext domain, and the processed data tend to lose significant magnitude relation. Therefore, in this section, we only focus on investigation of the comparison protocols without OPE. We will present some explanations on OPE in the Appendix "Key terminology and definitions."

6. Summary of key lessons learned/contributions of the research

After reviewing some benchmark protocols, we examine the differences between them by category, in order to obtain a comprehensive perspective on their security and performance.

6.1 Analysis of difference between comparison protocols categories

To conduct a comprehensive comparison of the security and performance levels of the protocols, we use the indicators as in Table 16:

The first two indicators reflect the security level of the protocols, and the last three indicators reflect the performance of those protocols. And the advantages and disadvantages of comparison protocols can be shown in Fig. 15. The red line denotes comparison protocols for CDO, the green line denotes comparison protocols for CCS with a trusty part, and the black line denotes comparison protocols for CCS without trusty part. The five sector blocks represent five different evaluation indicators. When the colored line shown in the sector, this type of protocol is expressed as YES in this indicator.

6.2 Achievements of the research

By conducting a comprehensive comparison of the benchmark protocols by categories, we learn that:

- Comparison protocols for CDO are of high security level. However, the data owners have to take on big workload/therefore, they are not efficient when the data volume is large.
- Comparison protocols for CCS with a trusty part can reduce the burden upon data owners by outsourcing heavy tasks to cloud servers. They can

Table 16 The indicators for comparison protocols categories.

	Indicator	Annotation
Security	No private key leaked	Is the DO's private key only kept by DO?
	Infrequent communication with DOs	Do the participants not need to communicate with DO frequently?
Performance	The demand for encryption is low	In addition to ensuring security, is there no other special requirement for the encryption algorithm?
	DOs do not take extra work	Do data owners need to design a unique function to complete the comparison?
	Multi-data sorting efficiently	Whether the protocol can sort multi-data in one round or not?

Source: author.

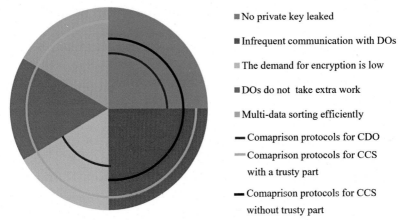

■ No private key leaked

■ Infrequent communication with DOs

■ The demand for encryption is low

■ DOs do not take extra work

■ Multi-data sorting efficiently

— Comaprison protocols for CDO

— Comaprison protocols for CCS
with a trusty part

— Comaprison protocols for CCS
without trusty part

Fig. 15 The advantages and disadvantages of the comparison protocols. *Source: author.*

efficiently make a comparison of the huge amount of data. However, the efficiency is achieved by data owners surrendering their own private keys to a trusty cloud server. This practice sacrifices the security of the protocols. Therefore, those protocols are not secure enough when being put into practice.

- Comparison protocols for CCS without trusty part have high requirements for the encryption algorithms. Those ciphertexts could be incompletely decrypted into comparable intermediate results, without leaking the original data. Although data owners need to design a suitable great than function to make comparison on the encrypted data, this workload is much less than processing data by themselves. As the data owners' private keys have not been compromised, the security of comparison protocols for CCS without trusty part is comparatively higher.

Data comparison is a part of data processing. Researchers need to choose the appropriate protocol according to the specific data processing mode and scenarios. Different application scenario has different requirements for security and efficiency level. The systematic investigation provides reference for selecting the most suitable protocol.

7. Research directions in the field

Nowadays, as cloud servers have been widely used, secure comparison protocols for CCS have a wider range of applications than comparison protocols for CDO. Existing data comparison protocols for CCS have their own advantages and disadvantages. To construct or select a most suitable protocol, the following may be taken into consideration: For security purpose, the protocols should be structured without a third part. And in order to reduce the data owners' computing burdens, the protocols should prevent data owners from undertaking too much workload. Moreover, the cloud servers should be able to provide service to all users, including users who do not have computer science technical background. The comparison protocols should be independent of data processing to get a wider range of applications.

On our research agenda, the next goal is to propose a privacy-preserving comparison protocol with proxy re-encryption, which can conduct comparison over encrypted data work, without leaking privacy of data owners.

8. Conclusion

In cloud computing, the privacy of the data owners is subject to the curiosity of many participants. Safe and effective comparison protocols are a step in guaranteeing data security, in addition to encryption step. In comparison protocols, getting the magnitude relation between encrypted data has become a crucial point of operation, where the data privacy is the most vulnerable. In former research, as different cryptosystems were used in different schemes, there emerged a lot of comparison protocols. After systematic investigation, we establish a classification system of the protocols, and reviewed benchmark protocols in each category within the system. Our system provide reference to constructing or selecting the most suitable protocols and also reveal where to improve the existing protocols. Future research can focus on how to constructing a comparison protocol maximize the strength, while minimizing the disadvantages of the existing ones: a protocol for CCS without leaking data owners' private keys. This protocol should also reduce the workload of data owners by adopting a simple great than function.

Acknowledgments

This work is supported by the National Key R&D Program of China Grant 2018YFB1003205; by the National Natural Science Foundation of China (No. U1836208, No. U1536206, No. U1836110, No. 61602253, No. 61672294); by the Jiangsu Basic Research Programs-Natural Science Foundation (No. BK20181407); by the Priority Academic Program Development of Jiangsu Higher Education Institutions (PAPD) fund; by the Collaborative Innovation Center of Atmospheric Environment and Equipment Technology, China (CICAEET) fund.

Appendix: Key terminology and definitions

Secure Multiparty Computation (SMC)—SMC is a subfield of cryptography. It aims at creating methods for parties to jointly compute a function over their inputs, while keeping those inputs private. Unlike traditional cryptographic tasks, where the adversary is outside the system of participants, the adversary in SMC controls actual participants. For security consideration, the participant's inputs should be confidential to anyone else.

Secure computation was formally introduced as secure two-party computation (2PC) in 1982 for the so-called Millionaires' Problem, and in generality in 1986 by Andrew Yao. SMC is considered as a practical solution to various real-life problems, especially the problems only require linear sharing of the secrets and local operations on the shares with not much interactions among the parties, such as distributed voting, private bidding and auctions, sharing of signature or decryption functions and private information retrieval.

Yao's millionaire problem—It is a classic problem in secure multiparty computation, which was introduced in 1982 by Andrew Yao [3]. The problem discusses two millionaires, Alice and Bob, who are interested in knowing which of them is richer without revealing their actual wealth. This problem is analogous to a more general problem where there are two numbers a and b and the goal is to solve the inequality $a \geq b$ without revealing the actual values of a and b. The Millionaires' Problem is an important problem in cryptography. The first solution was proposed by Yao himself [3], and many solutions have been proposed by other researchers.

Cryptography—It is the practice and study of techniques for secure communication in the presence of third parties called adversaries. It is about constructing and analyzing protocols that prevent third parties or the public from reading private messages. Modern cryptography exists at the intersection of the disciplines of mathematics, computer science, electrical engineering, communication science, and physics.

Homomorphic encryption—It is a special subclass of public key encryption scheme. It is a form of encryption that allows computation on ciphertexts, generating an encrypted result which, when decrypted, matches the result of the operations as if they had been performed on the plaintext. Homomorphic encryption can be used for secure outsourced computation.

Scale-Invariant Feature Transform (SIFT)—SIFT is an algorithm in computer vision to detect and describe local features in images. It is a feature that is widely used in image processing. The processes of SIFT include Difference of Gaussians (DoG) Space Generation, Keypoints Detection, and Feature Description.

1. DoG space generation

In the early literatures related to image processing, the images with different scales are represented by image pyramid as shown in Fig. 16. An image pyramid is a set of results obtained from the same image at different resolutions. Its generation process generally consists of two steps:

• Smoothing the original image.
• Downsampling the processed image.

Gaussian kernels are the only kernels that can generate multiscale space. The original image passes through Gaussian smoothing filters with different kernel parameters $\sigma's$ can output a set of images in different scales. The Gaussian kernel is shown as follows.

$$G(x, y, \sigma) = \frac{1}{2\pi\sigma^2} \, exp\left(-\frac{(M/2 - x)^2 + (N/2 - y)^2}{2\sigma^2}\right)$$

where M denotes the length of the kernel and N denotes the height of the kernel.

After that the Gaussian Scale Space can be obtained as follows.

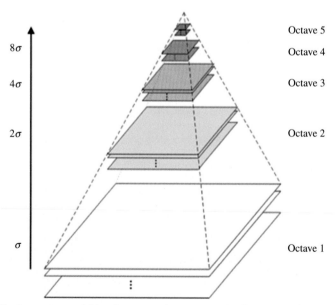

Fig. 16 The image pyramid by downsampling. *Source: author.*

$$L(x, y, \sigma) = G(x, y, \sigma) * I(x, y)$$

The downsampling will shrink the images (in general, the downsampling in SIFT will reduce the image's length and width to half its original length and width). The image pyramid with different scales can be obtained as Fig. 17. The DoG Scale Spaces can be obtained by taking the difference between two adjacent Gaussian Scale Spaces.

$$D(x, y, \sigma) = L(x, y, k\sigma) - L(x, y, \sigma)$$

And, the DoG can be shown as Fig. 15. DoG is an approximation of LoG (Laplacian of Gaussian). The LoG is one of the first and also most common blob detecting method. However, the calculation of LoG is complicated. Lowe succeeded in approximating LoG with DoG. The position of extreme point is same. And DoG is much simpler.

2. The key points detection

If a point is considered as a key point, it needs to be compared with eight neighbors in the same layer and nine neighbors in the upper and lower layers as shown in Fig. 18. It is the largest or smallest of the 27 points in order to be considered as the key point.

3. Feature description

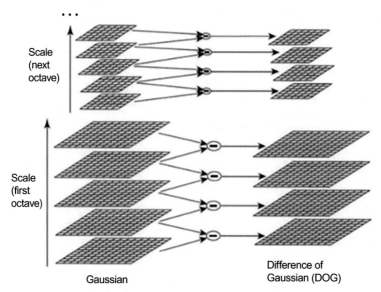

Fig. 17 The generation of DoG Space pyramid. *Source: author.*

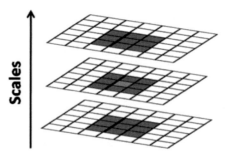

Fig. 18 The keypoints detection. *Source: author.*

The SIFT feature is the description of the gradient magnitude and gradient direction around the key points. First, take the pixels of 16×16 centered on the key point. Second, allocate them to 4×4 blocks. Third, divide 360° into eight directions on average, as 0°, 45°, and so on. Fourth, calculate the gradient magnitude of each direction in each block as shown in Fig. 19. Finally, all the gradient magnitudes are expressed in one vector in order, and the $4 \times 4 \times 8 = 128$-dimensional SIFT feature is obtained.

OPE Order-preserving Encryption (OPE)—OPE is a deterministic encryption scheme whose encryption function preserves numerical ordering of the plaintexts. OPE allows comparison operations to be directly applied on encrypted data, without decrypting the operands. Thus, equality and range queries as well as the MAX, MIN, and COUNT queries can be directly processed over encrypted data. Similarly, GROUP BY and ORDER BY operations can also be applied. Only when applying SUM or AVG to a group do the values need to be decrypted.

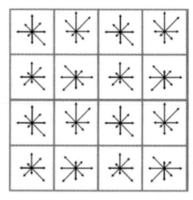

Fig. 19 The feature description. *Source: author.*

References

[1] Q. Lu, Y. Xiong, X. Gong, W. Huang, Secure collaborative outsourced data mining with multi-owner in cloud computing, in: IEEE International Conference on Trust, Security and Privacy in Computing and Communications, IEEE, 2012, pp. 100–108.

[2] L. Zhang, Y. Zheng, J. Weng, C. Wang, Z. Shan, K. Ren, You can access but you cannot leak: defending against illegal content redistribution in encrypted cloud media center, IEEE Trans. Depend. Secure Comput. (2018).

[3] A.C. Yao, Protocols for secure computations, in: Proceedings of the 23rd Annual IEEE Symposium on Foundations of Computer Science, Chicago, 1984, pp. 160–164.

[4] M.B. Karimi, A. Isazadeh, A.M. Rahmani, QoS-aware service composition in cloud computing using data mining techniques and genetic algorithm, J. Supercomput. 73 (2017) 1–29.

[5] D. Searson, Data mining and machine learning in e-science central using WEKA, Res. Microbiol. 159 (2015) 406–414.

[6] C. Sureshkumar, K. Iyakutty, Semantic cluster based classification for data leakage detection for the cloud security, Int. J. Comput. Appl. 110 (2015) 19–22.

[7] Y. Rahulamathavan, R. Phan, S. Veluru, K. Cumanan, M. Rajarajan, Privacy-preserving multi-class support vector machine for outsourcing the data classification in cloud, IEEE Trans. Depend. Secure Comput. 11 (2014) 467–479.

[8] Z. Xia, X. Ma, Z. Shen, X. Sun, N. Xiong, Secure image LBP feature extraction in cloud-based smart campus, IEEE Access 6 (2018) 30392–30401.

[9] Z. Abduljabbar, H. Jin, D. Zou, Z. Hussien, An efficient and robust one-time message authentication code scheme using feature extraction of iris in cloud computing, in: International Conference on Cloud Computing and Internet of Things, Changchun, 2015, pp. 22–25.

[10] J. Guo, H. Prasetyo, Content-based image retrieval using features extracted from halftoning-based block truncation coding, IEEE Trans. Image Process. 24 (2015) 1010–1024.

[11] Z. Xia, X. Wang, L. Zhang, Z. Qin, X. Sun, K. Ren, A privacy-preserving and copy-deterrence content-based image retrieval scheme in cloud computing, IEEE Trans. Inf. Foren. Secur. 11 (2016) 2594–2608.

[12] C.N. Yang, Z.X. Yeh, S.C. Hsu, Z. Fu, Enhancing security for two-party comparison over encrypted data, in: International Conference on Systems and Informatics, Shanghai, 2017, pp. 439–443.

[13] A. Maitra, G. Paul, A.K. Pal, Secure two-party quantum computation for non-rational and rational settings, arXiv (2016). 1504.019745v5.

[14] N. Emmadi, P. Gauravaram, H. Narumanchi, H. Syed, Updates on sorting of fully homomorphic encrypted data, in: International Conference on Cloud Computing Research and Innovation, Singapore, 2015, pp. 19–24.

[15] A. Chatterjee, M. Kaushal, I. Sengupta, Accelerating sorting of fully homomorphic encrypted data, in: Progress in Cryptology, INDOCRYPT 2013, Springer International Publishing, 2013, pp. 262–273.

[16] G.S. Cetin, Y. Doroz, B. Sunar, E. Savas, Depth optimized efficient homomorphic sorting, in: Progress in Cryptology, LATINCRYPT, 2015, pp. 61–80.

[17] M. Fischlin, A Cost-Effective Pay-per-Multiplication Comparison Method for Millionaires, Topics in Cryptology, CT-RSA 2001, Springer, Berlin Heidelberg, 2001, pp. 457–471.

[18] M. Atallah, W. Du, A Multi-Dimensional Yao's Millionaire Protocol, Purdue University, West Lafayette, USA, 2001.

[19] H. Jia, Q. Wen, T. Song, F. Gao, Quantum protocol for millionaire problem, Opt. Commun. 284 (2011) 545–549.

[20] S. Li, D. Wang, Y. Dai, P. Luo, Symmetric cryptographic solution to Yao's millionaires' problem and an evaluation of secure multiparty computations, Inform. Sci. Int. J. 178 (2008) 244–255.

[21] M. Nateghizad, Z. Erkin, R. Lagendijk, An efficient privacy-preserving comparison protocol in smart metering systems, EURASIP J. Inf. Secur. 2016 (2016) 11–19.

[22] M. Rao, M. Laxmaiah, Secure mining of association rules over horizontally partitioned data in data mining, Global J. Comp. Sci. Technol. 14 (2010) 84–88.

[23] S. Samet, A. Miri, Privacy-preserving classification and clustering using secure multi-party computation, in: Proceedings of Annual Acm Symposium on the Theory of Computing, 2, 2006, pp. 927–938.

[24] H. Lin, W. Tzeng, An efficient solution to the millionaires' problem based on homo-morphic encryption, in: Lecture Notes in Computer Science, 3531, Springer, Berlin, Heidelberg-Verlag, 2005, pp. 456–466.

[25] S. Hu, Q. Wang, J. Wang, Z. Qin, K. Ren, Securing SIFT: privacy-preserving out-sourcing computation of feature extractions over encrypted image data, IEEE Trans. Image Process. 25 (2016) 3411–3425.

[26] A. Kumar, A. Gupta, Some efficient solutions to Yao's millionaire problem, Comput. Sci. (2013). arXiv:1310.8063.

[27] C.Y. Hsu, C.S. Lu, S.C. Pei, Image feature extraction in encrypted domain with privacy-preserving SIFT, IEEE Trans. Image Process. 21 (2011) 4593–4607.

[28] R. Bellafqira, G. Coatrieux, D. Bouslimi, G. Quellec, M. Cozic, Secured outsourced content based image retrieval based on encrypted signatures extracted from homomorphically encrypted images, arXiv (2017). 1704.00457v1.

[29] L. Jiang, C. Xu, X. Wang, B. Luo, H. Wang, Secure outsourcing SIFT: efficient and privacy-preserving image feature extraction in the encrypted domain, IEEE Trans. Depend. Secure Comput. 14 (2015) 1–15.

[30] Q. Wang, Y. Luo, L. Huang, Privacy-preserving protocols for finding the convex hulls, in: 2008 Third International Conference on Availability, Reliability and Security, Barcelona, 2008, pp. 727–732.

[31] P. Kaghazgaran, B. Sadeghyan, Secure two party comparison over encrypted data, in: 2011 World Congress on Information and Communication Technologies, Mumbai, 2011, pp. 1123–1126.

[32] W. Zhao, M. He, Improvement of a secure convex hull two-party computation protocol, J. Converg. Inf. Technol. 5 (2013) 24–30.

[33] Z. Qin, J. Yan, K. Ren, C. Chen, Privacy-preserving outsourcing of image global fea-ture detection, in: 2014 IEEE Global Communications Conference, Austin, 2014, pp. 710–715.

[34] Q. Wang, S. Hu, K. Ren, J. Wang, Z. Wang, Catch me in the dark: effective privacy-preserving outsourcing of feature extractions over image data, in: IEEE INFOCOM 2016—The 35th Annual IEEE International Conference on Computer Communications, San Francisco, 2016, pp. 1–9.

[35] A. Boldyreva, N. Chenette, Y. Lee, A. O'Neill, Order-preserving symmetric encryp-tion, in: International Conference on Advances in Cryptology-Eurocrypt, DBLP, 2009.

[36] A. Boldyreva, N. Chenette, A. O'Neill, Order-preserving encryption revisited: improved security analysis and alternative solutions, in: Annual Cryptology Conference, Springer, Berlin, Heidelberg, 2011.

[37] Z. Qin, J. Yan, K. Ren, C. Chen, C. Wang, Towards efficient privacy-preserving image feature extraction in cloud computing, in: Proceedings of the 22nd ACM International Conference on Multimedia, 2014, pp. 497–506.

About the authors

Mr. Leqi Jiang is a Ph.D. Candidate in School of Computer and Software, Nanjing University of Information Science & Technology. He obtained his M.S. degree in Software Engineering from Nanchang Hangkong University, China, in 2016. His research interests include encrypted image processing, data security in cloud.

Prof. Zhihua Xia obtained his B.S. degree from Hunan City University, China and Ph. D. degree in computer science and technology from Hunan University, China, in 2006 and 2011, respectively. He works as an associate professor in School of Computer & Software, Nanjing University of Information Science & Technology. In recent years, he has published more than 50 papers in international journals as first author. He authored one invention patent. His papers were cited 1150+ times in Web of Science, 2000+ times in Google Scholar. Six papers became ESI highly cited papers. Some of his research results are adopted by the national security department.

Prof. Xingming Sun obtained his B.S. degree in Mathematics from Hunan Normal University, China, in 1984, and M.S. degree in Computing Science from Dalian University of Science and Technology, China, in 1988, and Ph.D. in Computing Science from Fudan University, China, in 2001. He conducted post-doctoral research at National University of Defense Technology. From December 2006 to December 2007, he was appointed as a government-sponsored visiting scholar to University College London. From March 2008 to February 2010, he was employed as a visiting professor by University of Warwick in the United Kingdom. He is currently the Dean of the School of Computer and Software, Nanjing University of Information Engineering and the Director of Jiangsu Network Monitoring Engineering Research Center. He is also a professor in China–USA Computer Research Center, China. He has also served as Member of the expert group of the National Natural Science Foundation of China (NSFC), Deputy Director of Huazhong Evaluation Center in China Information Security Evaluation Center, Member of National Expert Committee on Digital Watermarking and Information Hiding, Member of Specialized Committee on Theoretical Computer of China Computer Society. He is the reviewer of more than 10 international journals. He is the general chair of ICCCS (International Conference of Cloud Computing and Security) 2015, 2016, 2017, and 2018. He is a recipient of Science and Technology Progress Award, and a senior member of IEEE. His research interests include network and information security, digital watermarking, and data security in cloud.

CHAPTER FOUR

Provably secure verifier-based password authenticated key exchange based on lattices

Jinxia Yu[a], Huanhuan Lian[a], Zongqu Zhao[a], Yongli Tang[a], and Xiaojun Wang[b]

[a]Henan Polytechnic University, Jiaozuo, China
[b]Dublin City University, Dublin, Ireland

Contents

Advances in Computers, Volume 120
ISSN 0065-2458
https://doi.org/10.1016/bs.adcom.2020.09.003

Abstract

Verifier-based Password Authenticated Key Exchange (VPAKE) protocol enables users to generate a session key over insecure channels, which can limit the impact of server's information leakage. However, most existing VPAKE protocols are based on the *integer factorization problem* and the *discrete logarithm problem*; they cannot resist attack by quantum computers. In this chapter, we propose a new VPAKE protocol based on lattices. The protocol is constructed by using Chosen-Ciphertext Attacks (CCA) secure public-key encryption scheme, which is based on the *learning with errors problem* and an associated approximate smooth projective hash. Furthermore, this protocol uses a new randomized password hashing scheme based on lattices. This scheme enables ASCII-based passwords and a zero-knowledge password policy check; it allows users to prove the compliance of their password without revealing any information. Meanwhile, through explicit mutual authentication between the users and the servers, the protocol can resist undetectable online dictionary attacks. We then prove the security of this protocol. Our new protocol only involves three-round interactions with mutual explicit authentication. In addition, it avoids vulnerability of cryptosystem based on the *integer factorization problem*, and it is robust against quantum attacks.

Abbreviations

AKE	authenticated key exchange
ASCII	American standard code for information
ASPH	approximate smooth projective hash
CA	certification authority
CCA	chosen–ciphertext attacks
CRS	common reference string
ECC	elliptic-curve cryptography
KEM	key encapsulation mechanism
LWE	learning with error
NIST	National Institute of Standards and Technology
PAKE	password authenticated key exchange
PKE	public-key encryption
PPC	password policy check
QROM	quantum random oracle model
RO	random oracle
ROM	random oracle model
SIS	short integer solution
SPH	smooth projective hash
VPAKE	verifier-based password authenticated key exchange
ZKPPC	zero-knowledge password policy check

1. Introduction

Nowadays, people exchange information online through public channels. These channels are open and insecure. For realizing secure communication through insecure channels, a series of cryptographic systems have been invented and adopted. In these cryptographic systems, keys are to be encrypted and decrypted, and the process of key exchange is crucial.

Key exchange is a process in which two or more parties establish shared secret key for cryptographic use by a series of negotiation. Authenticated Key Exchange (AKE) is one of the most fundamental and widely used cryptographic primitives. It authenticates the identities of the users with pre-shared information, and establishes a high-entropy and secret session key over an insecure communication network. Usually, the shared information can be either a high-entropy cryptographic key (e.g., a secret key for symmetric-key encryption, or a public key for digital signature) or a low-entropy password. As the former requires additional public-key infrastructure, so most of such AKE protocols are easy to be put into practice. The low-entropy password is easy to remember and operate. Password Authenticated Key Exchange (PAKE) can avoid complex key management. Therefore, PAKE are getting more attention from researchers recently.

PAKE allows two users to authenticate each other with shared low-entropy passwords, and to establish a session key. Bellovin and Merritt [1] initiated research in this direction and designed the first PAKE protocol. Later, many PAKE protocols had been proposed. However, in most of the existing protocols, the servers are at risk of being compromised. In order to limit the damage by server compromise, protocols with a verifier-based model can be employed. In a verifier-based protocol, the server only stores a verifier, instead of the password itself, so server compromise does not directly leak the passwords.

In the existing literature, most verifier-based PAKE (VPAKE) protocols were based on traditional mathematical problems, such as *integer factorization problem* and *discrete logarithm problem*; they are insecure against attack by quantum computers. In recent years, Lattices emerges as a new powerful mathematical platform. A variety of cryptographic primitives have been built on this platform. Starting from the study by Shor [2], lattices have been used to construct public-key encryption, digital signatures, key exchange, etc. Lattice-based cryptography is attractive to researchers: it is simple, and efficient highly parallelized algorithm; it mainly involves the linear computation of the vectors and the matrix of small integer. Therefore, we attempt to construct a verifier-based PAKE protocol from lattices.

This chapter presents the development of authenticated key exchange, and the structures of PAKE. We propose a new verifier-based PAKE on lattice. Our contributions include: (1) we combine a randomized password hashing scheme based on the hardness of the Short Integer Solution (SIS) problem and a SIS-based Zero-knowledge Password Policy Check

(ZKPPC) to ensure the password security. (2) We construct a VPAKE by using Chosen-Ciphertext Attacks (CCA) secure public-key encryption scheme, which is based on the learning with errors problem and an associated approximate smooth projective hash.

The remainder of this chapter is as follows. Section 2 presents the classification of AKE and the progress of PAKE; Section 3 briefs major component technology in VPAKE; Section 4 presents the background of lattice; Section 5 demonstrates our new PAKE protocol based on lattices; Section 6 presents security model and security proof of our new protocol; Section 8 summarizes key lessons learned and research directions; Section 9 is the conclusion of our research.

2. Overview of AKE

2.1 The significance of AKE and its classification

With the rapid development of quantum computing, the security of traditional public-key cryptography is facing big challenges. At present, the security of public-key encryption schemes are mostly relying on the difficulties of *integer factorization problem* and *discrete logarithm problem*. These cryptography schemes were secure, because these problems can be solved by traditional computers only in exponential time or sub-exponential time. However, Shor's algorithm [2] showed that all public-key cryptography schemes based on the difficulties of *integer factorization problem* and *discrete logarithm problem*, such as RSA, ElGamal and ECC, etc., could be attacked by quantum computer in polynomial time. Therefore, in order to avoid the security risk brought by quantum computing, cryptographers have begun designing post-quantum cryptography schemes.

As an important theoretical basis for designing the quantum secure public-key cryptosystem, lattice-based cryptography is developing fast in recent years. Lattice-based cryptography can combat quantum and sub-exponential attack. They are efficient, and can conduct parallel implementations. They have potential to be put into practice. Some difficult problems based lattice are immune to the quantum attack, such as shortest vector problem, Learning with Error (LWE) problem, etc. These difficult problems can provide strong theoretical guarantee for cryptosystem in the post-quantum scenario. Starting from the study by Ajtai [3], lattices have been used to construct one-way functions and collision-resistant hash functions [3–5], signatures [6–8], public-key encryption [9–11], identity-based encryption schemes [12–15], key exchange protocol [16–18], trapdoor functions [12,19] and fully homomorphic encryption [20–23].

Key exchange is the core of modern cryptosystems. It is the process where two or more parties establish a common session key through a series of negotiations, to reach the goal of secure communication. The key exchange technology mainly involves two aspects: one is the key exchange based on symmetric cipher; the other is the key exchange based on asymmetric cipher (public-key cryptography). Key exchange [24] was first proposed by Diffie-Hellman in 1976. The protocol was based on the difficulty assumption of the discrete logarithm problem. It could only be used to exchange keys, but not for encryption and decryption. The Diffie-Hellman key exchange protocol was secure under passive adversaries, but it was insecure under active adversaries, who can fully control all communications. Because the protocol was in lack of the identity authentication of both parties, it could not resist man-in-the-middle attacks and impersonation attack. Therefore, developing AKE, which can overcome active adversaries, has become a significant task.

Depending on the ways of authentication, AKE can be classified as certificate-based AKE, identity-based AKE, certificate-less AKE, and password-based AKE. (see Table 1).

Table 1 Classification of AKE.

Type	Characteristics
Certificate-based AKE	Entity identity authentication based on certificate is the first authentication mechanism based on certificate. It was established with the idea of public-key cryptography. In the certificate-based authentication mechanism, there is a public trust center-Certification Authority (CA). The public center is responsible for the certificate issuance, renewal and revocation
Identity-based AKE	In the initialization phase, the key generation center generates its own private key, then generates the user's private key according to the user's identity, and sends the private key to the user through the secret channel. One of the advantages of the identity-based authentication is that the user's public key has a common correspondence with its own identity
Certificate-less AKE	Certificate-less AKE protocol attempts to provide complete privacy protection when a temporary key is compromised, or when there are some active attacks. These protocols mostly adopt the bilinear pairing method
Password-based AKE	Password-based AKE enables participants to authenticate others with pre-shared low-entropy password, and then establishes final session keys on public networks

Source: author.

2.1.1 Password-based AKE (PAKE)

The most widely used authentication method is password authentication, as the password is easy to remember, and additional storage devices are not required.

PAKE protocol can eliminate dependency on public-key infrastructure and security hardware, so it improves the convenience of the key exchange system. In a PAKE protocol, a secret value is pre-shared between the user and the server. This information is the password for the user; as for the server, this information may be a password or the transformation of the password. The server uses on this information to authenticate the user's identity, and to conduct key exchange. In this chapter, we focus on the PAKE protocol.

2.2 The development of PAKE protocol

As passwords are widely used, a lot of research had been conducted focusing on designing PAKE protocols. A secure PAKE should be able to resist off-line dictionary attacks, in which the adversary tries to find the password using information obtained during previous protocol executions. It should also be able to limit the adversary to on-line attacks where the adversaries simply run the protocol with honest users using password trials. Early work [25] (see also [26]) established a "hybrid" model in which users shared public keys in addition to a password. In the "password-only" setting, the users and the servers were required to share a password only. Bellovin and Merritt [1] initiated research in this direction. It was not until several years later that some formal models for PAKE were developed [27–29], and provably secure PAKE protocols were shown in the random oracle/ideal models [27,28,30]. As pointed out in [31,32], it could be insecure to use a low-entropy password as a cryptographic key. Goldreich and Lindell [33] constructed the first PAKE protocol without random oracles. Their approach remains as the only one for the plain model, where there is no additional setup. However, their protocol was inefficient in terms of communication, computation, and round complexity. Goldreich-Lindell's protocol [33] also could not tolerate concurrent executions by the same party. There were many provably secure PAKE protocols based on various hardness assumptions proposed by researchers. These protocols mainly fall into two legions: one legion started from the work of Bellovin and Merritt [1], followed by a lot of study focusing on PAKE in the random oracle/ideal models. They aimed at achieving the highest possible level of performance [27,28,30,34]. The second legion dated back to the work of Katz, Ostrovsky and Yung [35]. After that, Gennaro and Lindell [36] constructed a generic PAKE framework (in the Common Reference String (CRS) model) based on Smooth Projective Hash (SPH) functions [37]. This legion of study focused on seeking more efficient PAKE in the standard model [8,12,31,38–41], which used a simple, low-bandwidth reconciliation technique to reach exact agreement. Zhang and Yu [32] constructed a new PAKE framework with a non-adaptive Approximate Smooth Projective Hash (ASPH). The projection function of such ASPH only depended on the projection key, which made this protocol more efficient.

Most of above PAKEs from lattices store passwords as plaintext on the server, as they assume that the server is secure. However, with the rapid development of technology, even the server could be compromised.

In order to reduce the damage caused by the server's information leakage, Bellovin and Merritt [42] proposed a Verifier-based Password Authentication Key Exchange protocol, which was known as the augmented encrypted key exchange protocol. In VPAKE protocols, the server stores a means, as a verifier, to verifier that the users used the correct password. These verifiers are usually hash values of passwords with a salt. If the server is compromised, this method can limit the impact of information leakage on the server side. Although it does not prevent off-line dictionary attacks, it forces the adversary to spend a lot of time to learn many passwords. This gives enough time to the users to renew their passwords. Some VPAKE protocols [43–50] were proposed with improved performance, but they still have their own disadvantages. The protocol in Ref. [44] was vulnerable to off-line dictionary attack [45]. The protocol in Ref. [46] was the first secure three-party VPAKE protocol, but it did not quality provable security.

The existing research on VPAKE has been based on *integer factorization problem* and *discrete logarithm problem*. Theoretically they are not secure in quantum computers environment. Moreover, due to the traditional smooth projective and underlying public-key encryption, the efficiency of existing VPAKE protocols were limited. Therefore, it is necessary to establish a new VPAKE protocol to fill these gaps. In this chapter, we establish a new VPAKE based on lattices, an operation on matrix-vector, integrating chosen-ciphertext attacks secure public encryption and new suitable approximate smooth projective hash function. This protocol can realize explicit mutual authentication. Our new VPAKE has advantages of being able to prevent server information leakage. It is also robust against online dictionary attack and quantum attack.

3. Major component technology in VPAKE

In this section, we brief some major component technology in VPAKE. We utilize them in constructing our new protocol. They are password hashing, Zero-knowledge Password Policy Check (PPC), Public-key Encryption (PKE) and Smooth Projective Hash (SPH). The role of each component technology in the process of VPAKE is illustrated in Fig. 1.

First, the user chooses the password and registers the authentication information remotely on the server. The actual password does not need to be sent. The zero-knowledge proof is used to ensure that the password selected by the user conforms to the password policy of the server. The password and the salt value are used to calculate the verification element by using a password hashing scheme, and the server stores the verification element. Next, after the user and the server get the required information, the process of key exchange is performed. In this process, the public-key encryption scheme and smooth projection hash function are used to perform information interaction, and finally the session key required by both parties is generated.

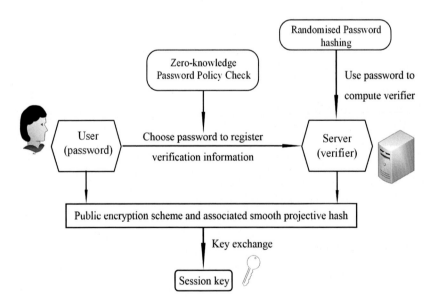

Fig. 1 The process of VPAKE. *Source: author.*

A password hashing scheme is a one-way process used to compute password hashing verification information. Zero-knowledge password policy check, users choose passwords on their own and register verification information on the server. They do not need to send password and they can check whether the password is compliant with the server password policies by zero-knowledge proof. Public encryption is used to encrypt the password, verifier and other information computed by the participants. In this way, leaking of important information from in the interactive process can be prevented. With properties of smoothness and correctness, smooth projective hash can help participants generate the same session key.

3.1 Randomized password hashing

The hashing scheme aims to compute password verification information that can be used later in VPAKE protocols. The user's password and a salt are the input of the password hashing scheme. The verifier is the output, which is stored on the server to authenticate the user's identity.

The users' login information cannot be used as the salt, because this may bring potential threats: first, it allows an attacker easily checking if a user uses the same password in two (corrupted) databases or in the same database during two different time periods; second, it allows creation of rainbow tables for specific logins, such as "root" or "administrator." As National Institute of Standards and Technology (NIST) recommends in Ref. [51], salts have to be unpredictable. They have to contain at least 128 randomly generated bits. This rule is executed and enforced by our definition (if the security parameter is set to 128 bits).

We use the randomized password hashing scheme as below:

$$\mathcal{H} = (\mathsf{Setup}, \mathsf{PPreSalt}, \mathsf{PPreHash}, \mathsf{PSalt}, \mathsf{PHash}).$$

- **Setup**(λ) takes as input security parameter λ and generates the public parameters pp;
- **PPreSalt**(pp) inputs pp, generates pre-hash salt s_P;
- **PPreHash**(pp, pw, s_P) takes as input pp, password pw and pre-hash salt s_P, outputs the pre-hash value P;
- **PSalt**(pp) takes as input pp and outputs s_H;
- **PHash**(pp, P, s_P, s_H) outputs the hash value h.

The password hashing scheme has five security properties: pre-image resistance, second pre-image resistance, pre-hash entropy preservation, entropy preservation and password hiding. When these security properties are in place, the adversary cannot obtain the password information unless they perform a brute-force attack, which allows them to find the pre-hash value or the password from the hash value stored in the server.

3.2 Zero-knowledge password policy check

A Password Policy Check (PPC) is an interactive protocol between a user and a server, where server's password policy $f = ((k_D, k_S, k_L, k_U), n_{min}, n_{max})$ and the public parameters $pp \leftarrow \mathsf{Setup}(\lambda)$ of a password hashing scheme. $\mathcal{H}$ is used as common inputs. If and only if $f(pw) = \textit{true}$ at the end of the PPC execution, the server accepts a hash value $\mathbf{h}$ for any password pw of the client's choice. A PPC protocol is a proof of knowledge of the password pw and randomness s_P, s_H used for hashing. A Zero-knowledge Password Policy Check (ZKPPC) protocol with zero-knowledge characteristic prevents the server from leaking any information about the password pw.

Let $\mathcal{H} = (\mathsf{Setup}, \mathsf{PPreSalt}, \mathsf{PPreHash}, \mathsf{PSalt}, \mathsf{PHash})$ be a randomized password hashing scheme. A ZKPPC protocol is a PPC protocol with zero-knowledge property between a prover (user) and verifier (server), such that: $f(pw) = \textit{true}$ and $\mathsf{PHash}(pp, P, s_P, s_H) = \mathbf{h}$, where $P \leftarrow \mathsf{PPreHash}$ (pp, pw, s_P).

3.3 Public-key encryption

A Public-key Encryption (PKE) with plaintext-space $\mathcal{P}$ consists of the following three algorithms:

- $(\textit{\textbf{pk}}, \textit{\textbf{sk}}) \leftarrow \mathsf{KeyGen}(1^\kappa)$: The key generation algorithm takes the security parameter κ as input, outputs a public key and a secret key (pk, sk).
- $c \leftarrow \mathsf{Enc}$ (pk, pw;r): the encryption algorithm takes pk and a plaintext $\textit{\textbf{pk}} \in \mathcal{P}$ as inputs, with an internal coin flipping r, outputs a ciphertext c.
- $pw \leftarrow \mathsf{Dec}(\mathrm{sk},c)$: the decryption algorithm takes sk and c as inputs, and outputs a plaintext pw or $\bot$.

Correctness

For all public key pk and secret key sk, any plaintext pw and c, the equation Dec $(\mathrm{sk},c) = pw$ holds with overwhelming probability.

Security

We consider the following game between a challenger $\mathcal{C}$ and an adversary $\mathcal{A}$.

3.4 Approximate smooth projective hash functions

Cramer and Shoup [37] first used the Smooth Projective Hash (SPH) functions to realize the public encryption that is robust against CCA. Later other study [31,36] applied this original definition in the field of PAKE protocols.

Let $\mathsf{PKE} = (\mathsf{KeyGen}, \mathsf{Enc}, \mathsf{Dec})$ be a secure public encryption scheme on lattices, $\mathcal{D}$ denote the plaintext space and C_{pk} denote the ciphertext space with respect to pk. Denote:

$$X = \left\{ (label, c) | label \in \{0, 1\}^*, c \in C_{pk} \right\};$$

$$L_m = \left\{ (C_{pk}, m) | \mathsf{Enc}(m; r) = C_{pk}, r \in \{0, 1\}^* \right\};$$

$$L = U_{m \in \mathcal{D}} L_m;$$

$$\overline{L}_m = \left\{ (C_{pk}, m) \in X | m = \mathsf{Dec}(C_{pk}) \right\};$$

$$\overline{L} = U_{m \in \mathcal{D}} \overline{L}_m;$$

An approximate SPH function is defined that: (1) sampling a hash key $hk \leftarrow_r K$, there is a Hash function cluster $\mathcal{H} = \{\mathsf{H}_{hk}\}_{hk \in K}$, where K denotes the hash key space; (2) there is a key projection function proj: $K \times C_{pk} \to HP$, where HP denotes the projection key space; (3) there is another Hash function cluster $\left\{ \mathsf{ProjH}_{hp} \right\}_{hp \in HP}$, where hp denotes the projection key. These functions satisfy the following properties:

ε-Approximate correctness
Ham(a,b) denotes the distance between a and b, where $a, b \in \{0, 1\}^n$. For any $x \in L$ and the associated projection key $hp = \mathsf{proj}(hk, C_{pk})$, there holds that
$$\Pr\left[\mathsf{Ham}\left(\mathsf{H}_{hk}(x), \mathsf{ProjH}_{hp}(hp, x, r) \geq \varepsilon \cdot n \right) \right] \leq \mathbf{negl}(n).$$

Smoothness
For any $x = (C_{pk}, m) \in X \backslash \overline{L}$, the two distribution $\{hp, \mathsf{H}_{hk}(x): hk \leftarrow HK; hp = \mathsf{proj}(hk, C_{pk})\}$ and $\{(hp, v): hk \leftarrow HK; hp = \mathsf{proj}(hk, C_{pk}); v \leftarrow \{0, 1\}^n\}$ have statistical distance negligible in n.

4. Theoretical background of lattice

In this section, we brief about lattice, on which our protocol is built. First, we explain notations and lattice theory; then we elaborate Short

Integer Solution (SIS) problem, on which randomized password hashing is based; and also LWE problem, on which the public encryption scheme is based.

4.1 Notations

The notations used in this chapter and the meaning of the variables in our study are detailed in Table 2.

4.2 Lattice theory

Based on the new cryptosystem, lattice theory has advantages of asymptotic efficiency, simple operation. It is also capable of parallel computing. In the quantum cryptography era, it has become a hot research spot.

Table 2 Notations description.

Symbol	Description
κ	Security parameter
$\mathbf{x}^T$	Row vectors, a transpose vector $\mathbf{x}$
$\mathbf{A}$	Matrices, sometimes identify a matrix with its ordered set of column vectors
a_i	Matrix $\mathbf{A}$'s i-th column vector
$\leftarrow_r$	Randomly choosing elements from some distribution
$\|\mathbf{R}\|$	The length of matrix $\mathbf{R}$, which is the norm of its longest column
$\widetilde{\mathbf{R}}$	The Gram-Schmidt orthogonalization of matrix $\mathbf{R}$
$s_1(\mathbf{R})$	The maximal singular value of matrix $\mathbf{R}$
$\mid$	The horizontal concatenation of matrices and/or vectors, e.g., $[\mathbf{B}\mid\mathbf{Bx}]$ denote the vertical concatenation as $[\mathbf{B};\mathbf{Bx}]$
$\|$	The (ordered) concatenation of vectors or matrices
$negl(n)$	A negligible function is an $f(n)$ such that $f(n) < (n^{-c})$ for every fixed constant c
Poly(n)	An unspecified function $f(n) = O(n^c)$ for some constant c
$\frac{1}{2}\sum_{d\in D}\|\mathcal{X}(d) - \mathcal{Y}(d)\|$	The statistical distance between two distributions $\mathcal{X}$ and $\mathcal{Y}$ (or two random variables having those distributions), viewed as functions over a countable domain D

Source: author.

We describe the definition of lattice, and other related definition and discrete Gaussian distribution, a core concept which lattice operation involves.

An m-dimensional full-rank lattice $\Lambda \subset \mathbb{R}^m$ is the set of all integral combinations of m linearly independent vectors $\mathbf{B} = (\mathbf{b}_1, \ldots, \mathbf{b}_m) \in \mathbb{R}^{m \times m}$, i.e., $\Lambda = \mathcal{L}(\mathbf{B}) = \{\sum_{i=1}^{m} x_i \mathbf{b}_i : x_i \in \mathbb{Z}\}$.

The dual lattice

Λ^* is the dual lattice of Λ, is defined to be $\Lambda^* = \{\mathbf{x} \in \mathbb{R}^m : \forall \mathbf{v} \in \Lambda, \langle \mathbf{x}, \mathbf{v} \rangle \in \mathbb{Z}\}$.

Let $n \geq 1$ and modulus $q \geq 2$ be integers, where n is the main security parameter throughout this work, and all other parameters are implicitly functions of n. An m-dimensional lattice from the family is specified relative to the additive group $\mathbb{Z}_q^n$ by a parity check matrix $\mathbf{A} \in \mathbb{Z}_q^{n \times m}$. The associated lattice is defined as follows:

$$\Lambda^{\perp}(A) = \{x \in \mathbb{Z}^m : Ax = 0 \bmod q\}$$
$$\Lambda_u^{\perp}(A) = \{x \in \mathbb{Z}^m : Ax = u \bmod q\}$$
(1)

Note that $\Lambda_u^{\perp}(A)$ is a coset of $\Lambda^{\perp}(A)$.

For any $s > 0$, $\mathbf{x} \in \Lambda$, centring at $\mathbf{c} \in \mathbb{R}^m$, the Gaussian function over $\Lambda \subseteq \mathbb{Z}^m$ is defined as $\rho_{s, \mathbf{c}}(\mathbf{x}) = \exp(-\boldsymbol{\pi} \|\mathbf{x} - \mathbf{c}\|^2 / s^2)$. Let $\rho_{s, \mathbf{c}}(\Lambda) = \sum_{\mathbf{x} \in \Lambda} \rho_{s, \mathbf{c}}(\mathbf{x})$, for $\mathbf{y} \in \Lambda$, any $s > 0$, centring at $\mathbf{c} \in \mathbb{R}^m$, the discrete Gaussian distribution is defined as $D_{\Lambda, s, \mathbf{c}}(\mathbf{y}) = \frac{\rho_{s, \mathbf{c}}(\mathbf{y})}{\rho_{s, \mathbf{c}}(\Lambda)}$.

Lemma 1. *Let S be any basis of $\Lambda^{\perp}(A)$ for some $A \in \mathbb{Z}_q^{n \times m}$ whose columns generate $\mathbb{Z}_q^n$, let $u \in \mathbb{Z}_q^n$ be arbitrary, and $\sigma = \|\widetilde{S}\| \cdot \omega(\sqrt{\log n})$. We have* $\Pr_{\mathbf{e} \leftarrow D_{\Lambda_u^{\perp}(A), \sigma}} [\|\mathbf{e}\| > \sigma \cdot \sqrt{m}] \leq negl(n)$.

Micciancio and Goldwasser [52] showed that a full-rank set S in a lattice Λ can be converted into a basis T for Λ with an equally low Gram-Schmidt norm.

Lemma 2. *[52]: Let Λ be an m-dimensional lattice. There is a deterministic polynomial-time algorithm that, given an arbitrary basis of Λ and a full-rank set $S = \{s_1, s_2, \ldots, s_m\}$ in Λ, returns a basis T of Λ satisfying $\|\widetilde{T}\| \leq \|\widetilde{S}\|$ and $\|T\| \leq \|S\| \sqrt{m}/2$.*

4.3 Short Integer Solution (SIS) problem

The Short Integer Solution (SIS) problem was first introduced in Ref. [3], and had become the theoretical foundation for one-way and collision-resistant hash functions, identification schemes, digital signatures, and other primitives (but not public-key encryption). The difficult problem of pass-word hashing scheme in our protocol is based on assumption of SIS problem.

The SIS problem can be seen as an average-case short-vector problem on a certain family of so-called "q-ary" m-dimensional integer lattices, namely, the lattices

$$\mathcal{L}^{\perp}(\mathbf{A}):=\left\{\mathbf{z} \in \mathbb{Z}^m : \mathbf{Az} = \mathbf{0} \in \mathbb{Z}_q^n\right\} \supseteq q\mathbb{Z}^m \tag{2}$$

Definition 1. (*SIS$_{n, q, \mathcal{B}, m}$*) [3,12] Given m uniformly random vectors $\mathbf{a_i} \in \mathbb{Z}_q^n$, forming the columns of a matrix $\mathbf{A} \in \mathbb{Z}_q^{n \times m}$, find a nonzero integer vector $\mathbf{z} \in \mathbb{Z}^m$ of norm $\|\mathbf{z}\| \leq \mathcal{B}$ such that

$$f_{\mathbf{A}}(\mathbf{z}):=\mathbf{Az} = \sum_i \mathbf{a_i} \cdot z_i = \mathbf{0} \in \mathbb{Z}_q^n \tag{3}$$

Lending the coding theory, here $\mathbf{A}$ acts as a "parity-check" (or more accurately, "arity-check") matrix that defines the lattice $\mathcal{L}^{\perp}(\mathbf{A})$. The SIS problem requires to find a sufficiently short nonzero vector in $\mathcal{L}^{\perp}(\mathbf{A})$, where $\mathbf{A}$ is chosen uniformly at random.

One can also consider an inhomogeneous version of the SIS problem, which is to find a short integer solution to $\mathbf{Az} = \mathbf{u} \in \mathbb{Z}_q^n$, where $\mathbf{A}$, $\mathbf{u}$ are uniformly random and independent. Note that, regardless of the norm constraint, the set of all solutions is the lattice coset $\mathcal{L}_{\mathbf{u}}^{\perp}(\mathbf{A}) = \mathbf{c} + \mathcal{L}^{\perp}(\mathbf{A})$, where $\mathbf{c} \in \mathbb{Z}^m$ is an arbitrary (not necessarily short) solution. These two problems, homogeneous and inhomogeneous, are essentially equivalent under certain parameters.

4.4 Learning with errors problem

The work [11] introduced the Learning with Errors (LWE) problem, which is the "encryption-enabling" analog of the SIS problem. These two problems are syntactically similar, and can be viewed seen as duals of each other. The difficult problem of public encryption scheme in our protocol is based on assumption of LWE problem. We describe LWE and its hardness as below:

Definition 2. (LWE). For any positive integers $n, q \in \mathbb{Z}$, real $\alpha > 0$ and vector $\mathbf{s} \in \mathbb{Z}_q^n$, define the distribution $A_{\mathbf{s},\alpha} = \{(\mathbf{a}, \mathbf{a}^t\mathbf{s} + e \bmod q) : \mathbf{a} \leftarrow_r \mathbb{Z}_q^n, e \leftarrow_r D_{\mathbb{Z}, \alpha q}\}$. For any m independent samples $(\mathbf{a}_1, b_1), \ldots, (\mathbf{a}_m, b_m)$ from $A_{\mathbf{s}, \alpha}$, we denote it in matrix form $(\mathbf{A}, \mathbf{b}) \in \mathbb{Z}_q^{n \times m} \times \mathbb{Z}_q^m$, where $\mathbf{A} = (\mathbf{a}_1, \ldots, \mathbf{a}_m)$ and $\mathbf{b} = (b_1, \ldots, b_m)^t$. If for uniformly random $\mathbf{s} \leftarrow_r \mathbb{Z}_q^n$ and given polynomial many samples, no PPT algorithm can recover **s** with non-negligible probability, we say that the $\boldsymbol{LWE}_{n,\,q,\,\alpha}$ problem is hard.

The decisional LWE problem, denoted $\textbf{distLWE}_{q,\,\chi}$ is asked to distinguish polynomial many samples from uniform. For certain parameters, the decisional LWE problem is polynomial equivalent to its search version, which is in turn known to be at least as hard as quantum approximating SIVP on n-dimensional lattices to within polynomial factors in the worst case [11].

5. Toward a new VPAKE based on lattice

In order to combat attack in quantum computer environment, we construct a new VPAKE protocol based on lattices. This protocol can provide mutual explicit authentication. We combine a randomized password hashing scheme based on the hardness of the SIS problem and a SIS-based ZKPPC to realize the password security and authentication.

Moreover, we construct a suitable ASPH function, which involves the language defined in this chapter; then we use the CCA-secure public-key encryption scheme based on the hardness of LWE problem and ASPH function, to obtain the first VPAKE from lattices.

In this section, we will first explain the construction of the randomized password hashing scheme and the CCA-secure public encryption, which are based on lattice. Then we explain the construction of a suitable ASPH function in the VPAKE protocol. After that, we present the process of our protocol.

5.1 Components in our VPAKE

➤ Randomized password hashing from lattices

This mechanism aims to compute some password verification information that can be used later in more advanced protocols (e.g., ZKPPC [53] and VPAKE [50]). In order to prevent off-line dictionary attacks, the computation process is randomized via pre-hash salt and hash salt. Our scheme $\mathcal{L}$ is based on lattice, defined as follows:

- $\mathcal{L}.\mathsf{Setup}(\lambda)$ takes as input security parameter λ and generates the public parameters $pp = (n, q, m, \mathbb{S}_P, \mathbb{S}_H, \mathbf{A}, \mathbf{B})$.
- $\mathcal{L}.\mathsf{PPreSalt}(pp)$ inputs pp., samples $\chi \leftarrow \mathcal{S}_{n_{\max}}$ and outputs $s_P = \chi$.

- $\mathcal{L}$.PPreHash(pp, pw, s_P) outputs the pre-hash value P. Compute encode$(pw) \in \{0,1\}^{8t}$ to get $\mathbf{e} \in \{0,1\}^{8n_{max}}$. Apply $T_{\chi,8}$ to get $\mathbf{e}' = T_{\chi,8}(\mathbf{e}) \in \{0,1\}^{8n_{max}}$ and output $P = \mathbf{e}'$.
- $\mathcal{L}$.PSalt(pp) samples $\mathbf{r} \leftarrow \{0,1\}^m$ and outputs $s_H = \mathbf{r}$.
- $\mathcal{L}$.PHash(pp, P, s_P, s_H) outputs the hash value h. Form.

$$\mathbf{e_0} = (\boldsymbol{bin}(\chi(1) - 1)|\ldots|\boldsymbol{bin}(\chi(n_{max}) - 1)) \in \{0,1\}^{n_{max}\lceil \log n_{max}\rceil} \qquad (4)$$

then form $\mathbf{x} = (\mathbf{e_0}\|\mathbf{e}') \in \{0,1\}^{n_{max}\lceil \log n_{max}\rceil + 8n_{max}}$ and output $\mathbf{h} = \mathbf{A} \cdot \mathbf{x} + \mathbf{B} \cdot \mathbf{r} \in \mathbb{Z}_q^n$.

Under the SIS assumption, the security of the randomized password hashing scheme follows five properties: pre-hash entropy preservation, entropy preservation, password hiding, pre-image resistance and second pre-image resistance.

Proof

The pre-hash salt $s_P = \chi$, if s_P is hidden, because it is randomly chosen, the min-entropy of P is larger than the min-entropy of pw. We first know that the pre-hash entropy preservation and entropy preservation properties stand. Second, the hash value $\mathbf{h} = \mathbf{A} \cdot \mathbf{x} + \mathbf{B} \cdot \mathbf{r} \bmod q$ has the statistical hiding property, so the password hiding property holds. Besides, the hash value has the computational binding property, if there are distinct pre-hash values P, P' that yield the same hash value $\mathbf{h}$, then these values break the property. Thus under the SIS assumption, the scheme has the second pre-image resistance property. Finally, over the randomness of matrix A, password pw and pre-hash salt s_P, except for a negligible probability (i.e., in the event one accidentally finds a solution to the SIS problem associated with matrix $\mathbf{A}$), vector $\mathbf{A} \cdot \mathbf{x}$ accepts at least 2^β values in $\mathbb{Z}_q^n$, where β is the min-entropy of the dictionary $\mathcal{D}$ from which pw is chosen. Therefore, even if $\mathbf{A} \cdot \mathbf{x} = \mathbf{h} - \mathbf{B} \cdot s_H \bmod q$ is given, to find $P = \mathbf{e}'$, one has to perform 2^β invocations of PPreHash which implies that the scheme satisfies the pre-image resistance property.

➤ CCA-secure Public-key encryption

The labeled public-key encryption is a CCA-secure public-key encryption scheme based on the hardness of the LWE problem. The PKE scheme PKE $=$ (KeyGen, Enc, Dec) is defined as follows:

Parameters. Let $n \in \mathbb{Z}$, a positive integer $m = m(n, \ell)$ and prime $q = q(n, \ell)$ be the parameters of the system. A Gaussian error parameter $\beta = \beta(n, \ell) \in (0, 1]$ that defines a distribution β.

- KeyGen(1^κ): Given the security parameter κ, choose a matrix $\mathbf{A}' \in \mathbb{Z}_q^{m \times (n+\ell+1)}$ with the trapdoor $\mathbf{T}$, compute $(\mathbf{A}', \mathbf{T}) \leftarrow \mathsf{TrapSamp}\left(1^m, 1^{n+\ell+1}, q\right)$. Return the public and secret key pair $(pk, sk) = (\mathbf{A}', \mathbf{T})$.
- Enc(pk,m;r): Given $pk = \mathbf{A}'$, $label \in \{0,1\}^*$ and message $\mathbf{w} \in \mathbb{Z}_q^\ell$, randomly choose $\mathbf{s} \leftarrow \mathbb{Z}_q^n$ and an error vector $\mathbf{x}' \leftarrow \overline{\Psi}_\beta^{mn}$. Finally, return the ciphertext.

$$\mathbf{y} = \mathbf{A}' \cdot \begin{pmatrix} \mathbf{s} \\ 1 \\ \mathbf{w} \end{pmatrix} + \mathbf{x}' \,(\bmod\, q) \tag{5}$$

- Dec(sk,c): Given $sk = \mathbf{T}$, compute

$$\begin{pmatrix} \mathbf{s} \\ a' \\ \mathbf{w} \end{pmatrix} \leftarrow \mathsf{BDDSolve}(\mathbf{T}, a\mathbf{y}) \tag{6}$$

for a = 1 to q-1 and if $a' = a$ then out $\mathbf{w}/a$ and stop, else try the next value of a.

Theorem 1. *Given* n, ℓ, m, q, β, *where* $m \geq 4(n+\ell)\log q$ *and* $\beta 1/(2 \cdot m^2 n \cdot (\sqrt{\log n}))$. *Then the CCA-secure encryption scheme assuming the hardness of* $\mathbf{distLWE}_{n,m,q,\beta}$.

➢ The associated approximate smooth projective hash function

We construct a suitable ASPH in our application. The public-key encryption scheme $\mathsf{PKE} = (\mathsf{KeyGen}, \mathsf{Enc}, \mathsf{Dec})$ is semantically secure from lattice. Let $\mathcal{P}$ be an efficiently recognizable plaintext space of $\mathcal{PKE}$, and let C_{pk} denote valid ciphertext space with respect to any public key pk. Let X denote a set and L denote a language with $L \subset X$.

$$X = \left\{ (label, c) \mid label \in \{0, 1\}^*, c \in C_{pk} \right\},$$
$$L_H = \left\{ (label, c) \mid \exists r, c = \mathsf{Enc}(pk, label, H; r) \right\},$$
$$L_{s,H} = \left\{ (label, c) \mid \exists P, \exists r, c = \mathsf{Enc}(pk, label, H; r) \wedge H = \mathsf{PHash}(pp, P, s_P, s_H) \right\},$$

where $s = (s_P, s_H)$ and some $\epsilon \in (0, 1/2)$.

There are efficient algorithms:

- HKGen(K): Sample a hash key $hk = (\mathbf{g}_1, ..., \mathbf{g}_\ell)$, where $\mathbf{g}_i \sim D_{\mathbb{Z}^m, \gamma}$.
- PKGen(hk): Compute hp. $= \mathsf{Proj}(\mathbf{g}_1, ..., \mathbf{g}_\ell) = (\mathbf{u}_1, ..., \mathbf{u}_\ell)$, $\mathbf{u}_i = \mathbf{B}^T \mathbf{e}_i$ where $\mathbf{B} = \mathbb{Z}_q^{m \times n}$.

- Hash(hk, L_H, c): Given all $\boldsymbol{hk} \leftarrow_r K$ and $c \in X$. First we compute $z_i = \mathbf{g}_i^T \left[\mathbf{y} - \mathbf{U} \cdot \begin{pmatrix} 1 \\ \mathbf{h} \end{pmatrix} \right] \in \mathbb{Z}_q$. Then treat each z_i as a number in $[-(q-1)/2$ $\ldots (q-1)/2]$. If $z_i = 0$, set $b_i \leftarrow_r \{0, 1\}$, else set $b_i = \begin{cases} 0 \text{ \textit{if} } z_i < 0 \\ 1 \text{ if } z_i > 0 \end{cases}$.

- ProjHash(hp, $L_{s_H, H}$, c, w, r): On input the projection key hp. = $(\mathbf{u}_1, \ldots, \mathbf{u}_\ell) \in S$, a ciphertext $(\boldsymbol{label}, \mathbf{y}) \in L$ and a witness $\mathbf{s} \in \mathbb{Z}_q^n$, compute $z_i' = \mathbf{u}_i^T \mathbf{s}$. Treat z_i' as a number in $[-(q-1)/2 \ldots (q-1)/2]$. If $z_i' = 0$, set $b_i' \leftarrow_r \{0, 1\}$, else set $b_i' = \begin{cases} 0 \text{ \textit{if} } z_i' < 0 \\ 1 \text{ if } z_i' > 0 \end{cases}$.

Approximate correctness: For $(\boldsymbol{label}, \mathbf{y}) \in L$, where y is a ciphertext. Note that y can be written as $\mathbf{y} = \mathbf{B}' \cdot \mathbf{s} + \mathbf{U} \cdot \begin{pmatrix} 1 \\ \mathbf{h} \end{pmatrix} + \mathbf{x}' \pmod q$ where $|\mathbf{x}'| \leq \beta q$ $\cdot \sqrt{mn}$. For each $i \in [\ell]$, we have that $z_i = \mathbf{g}_i^T \left(\mathbf{y} - \mathbf{U} \begin{pmatrix} 1 \\ \mathbf{h} \end{pmatrix} \right) = \mathbf{g}_i^T$ $(\mathbf{B}' \mathbf{s} + \mathbf{x}') = \mathbf{u}_i^T \mathbf{s} + \mathbf{g}_i^T \mathbf{x}'$. This means that $|z_i - z_i'| = |\mathbf{g}_i^T \mathbf{x}'| \leq \|\mathbf{e}_i\| \cdot \|\mathbf{x}'\| \leq \gamma \sqrt{mn} \cdot \beta q \sqrt{mn} \leq \varepsilon/2 \cdot q/4$ with the overwhelming probability. Each $\mathbf{u}_i$ is statistically close to uniform over $\mathbb{Z}_q^n$, we have that $z_i^T = \mathbf{u}_i^T \mathbf{s}$ is uniformly random. So we have the probability that $b_i \neq b_i'$ is at most $\varepsilon/2$. By a Chernoff bound, there holds $\Pr[\mathsf{Ham}(\mathsf{Hash}(\boldsymbol{hk}, L, c), \mathsf{ProjHash}(\boldsymbol{hp}, L, c, w, r)) \geq \varepsilon \cdot \ell] = \boldsymbol{negl}(\kappa)$.

Smoothness

For $(\boldsymbol{label}, \mathbf{y}) \in X \backslash L$, the value $\mathsf{Hash}(hk, L, c)$ is statistically indistinguishable from a random element, even when hp. is known.

Our ASPH notion involves the new language defined in the chapter. The projection function only depends on the hash key, and the encrypted message are correct hash value. These modifications allow us to limit the leakage of password.

5.2 Our VPAKE protocol

In our VPAKE protocol, the server S, which uses (id, h, r) stored from the ZKPPC-based registration protocol from lattice, can verify the identity of the client C, who only use his own pw. This protocol is constructed by modifying Benhamouda and Pointcheval's VPAKE framework [49].

Our protocol uses a randomized password hashing scheme $\mathcal{H}$ = (Setup, PPreSalt, PPreHash, PSalt, PHash). PKE = (KeyGen, Enc, Dec) is a

CCA-secure public-key encryption. The associated ϵ-approximate smooth projective hash $(K, \ell, \{\mathsf{Hash}_{hk} : X \rightarrow \{0, 1\}^{\ell}\}_{hk \in K}, HP, \mathsf{Proj} : K \rightarrow HP)$, hk, hp. denote hash key and projective key, K denotes the space of hash keys and HP denote the space of projective keys. Let the session key space be $\{0, 1\}^{\kappa}$, where κ is the security parameter. r_C denotes random value of client C and tk denotes temporary session key. Let $\mathsf{ECC} : \{0, 1\}^{\kappa} \rightarrow \{0, 1\}^{\ell}$ be an error-correcting code which can correct 2ϵ-fraction of errors, and ECC^{-1} be the decoding algorithm. Let $\boldsymbol{H_1}, \boldsymbol{H_2}$ be one-way hash functions. A description of our VPAKE is given in Fig. 2.

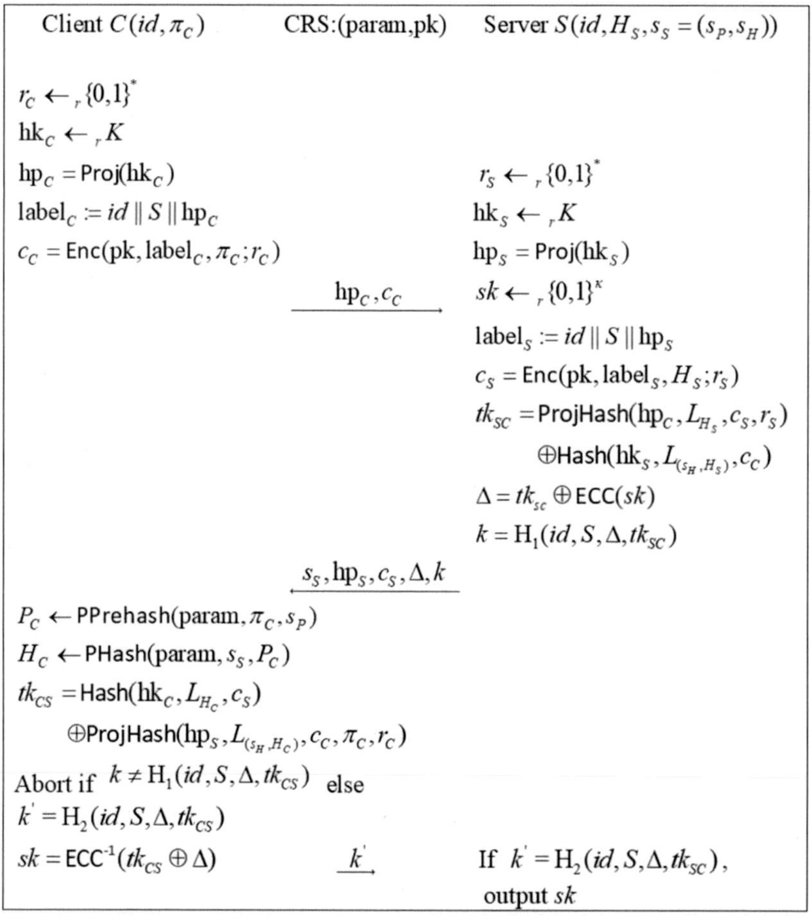

Client $C(id, \pi_C)$	CRS:(param,pk)	Server $S(id, H_S, s_S = (s_P, s_H))$

$r_C \leftarrow_r \{0,1\}^*$
$hk_C \leftarrow_r K$
$hp_C = \mathsf{Proj}(hk_C)$
$label_C := id \| S \| hp_C$
$c_C = \mathsf{Enc}(pk, label_C, \pi_C; r_C)$

$r_S \leftarrow_r \{0,1\}^*$
$hk_S \leftarrow_r K$
$hp_S = \mathsf{Proj}(hk_S)$

$$\xrightarrow{\quad hp_C, c_C \quad}$$

$sk \leftarrow_r \{0,1\}^{\kappa}$
$label_S := id \| S \| hp_S$
$c_S = \mathsf{Enc}(pk, label_S, H_S; r_S)$
$tk_{SC} = \mathsf{ProjHash}(hp_C, L_{H_s}, c_S, r_S)$
$\quad \oplus \mathsf{Hash}(hk_S, L_{(s_H, H_S)}, c_C)$
$\Delta = tk_{sc} \oplus \mathsf{ECC}(sk)$
$k = H_1(id, S, \Delta, tk_{SC})$

$$\xleftarrow{\quad s_S, hp_S, c_S, \Delta, k \quad}$$

$P_C \leftarrow \mathsf{PPrehash}(param, \pi_C, s_P)$
$H_C \leftarrow \mathsf{PHash}(param, s_S, P_C)$
$tk_{CS} = \mathsf{Hash}(hk_C, L_{H_C}, c_S)$
$\quad \oplus \mathsf{ProjHash}(hp_S, L_{(s_H, H_C)}, c_C, \pi_C, r_C)$
Abort if $k \neq H_1(id, S, \Delta, tk_{CS})$ else
$k' = H_2(id, S, \Delta, tk_{CS})$
$sk = \mathsf{ECC}^{-1}(tk_{CS} \oplus \Delta)$

$$\xrightarrow{\quad k' \quad}$$

If $k' = H_2(id, S, \Delta, tk_{SC})$,
output sk

Fig. 2 Our VPAKE with ASPH from lattices. *Source: author.*

5.2.1 Setup

We assume a CRS is established before any executions of the protocol. The CRS consists two parts: a public key pk of the CCA-secure encryption scheme, which can be generated by $KeyGen(1^{\kappa})$; and other public parameters. The client C holds password pw_C, which is mapped to π_C. The server S stores the hash value and salt of the client (s_S, H_S). The private information is the hash value $H_S = H_C$, and the server does not hold any word and the client holds a word $P_C = \mathsf{PPrehash}(\textbf{\textit{param}}, \pi_C) \in \{P| \exists \pi \in \{\textbf{0}, \textbf{1}\}^*,$ $\mathsf{PPrehash}(\textbf{\textit{param}}, \pi) = P \textbf{ and } \mathsf{PHash}(\textbf{\textit{param}}, P, s_S, \pi) = H_S\}$.

5.2.2 Protocol execution

We assume that client C uses π_C as its input and has already sent its login name id to server S, who picked the corresponding tuple (id, H_S, s_S) from its password database. Note that C can also act as initiator and send its id as part of its message, in which case S must act as a responder. First, C chooses random coins $r_C \leftarrow_r \{0, 1\}^*$ for encryption, a hash key $\textbf{\textit{hk}}_C \leftarrow_r K$ for ASPH, and computes the projection key $\textbf{\textit{hp}}_C = \textbf{\textit{Proj}}(\textbf{\textit{hk}}_C)$. Then the client sets $\textbf{\textit{label}}_C := id\|S\|\textbf{\textit{hp}}_C$, and it computes $c_C = \mathsf{Enc}(pk, \textbf{\textit{label}}_C, \pi_C; r_C)$ and sends $(\textbf{\textit{hp}}_C, c_C)$ to the server S.

Upon receiving $(\textbf{\textit{hp}}_C, c_C)$ from the client C, S chooses random coins $r_C \leftarrow_r \{0, 1\}^*$ for encryption, a hash key $\textbf{\textit{hk}}_C \leftarrow_r K$ for ASPH, and a random session key $sk \leftarrow_r \{0, 1\}^{\kappa}$ for some ℓ to be specified. It computes $\textbf{\textit{hp}}_S = \mathsf{Proj}(\textbf{\textit{hk}}_S)$, and defines $\textbf{\textit{label}}_S := id\|S\|\textbf{\textit{hp}}_S$, then server S computes $c_S = \mathsf{Enc}(pk, \textbf{\textit{label}}_S, H_S; r_S)$, $tk_{sc} = \mathsf{ProjHash}(\textbf{\textit{hp}}_C, L_{H_S}, c_S, r_S) \oplus \mathsf{Hash}$ $(\textbf{\textit{hk}}_S, L_{(s_H, H_S)}, c_C)$, and $\Delta = tk_{sc} \oplus \mathsf{ECC}(sk)$, $k = H_1(id, S, \Delta, tk_{SC})$. Finally, S sends the message $(s_S, \textbf{\textit{hp}}_S, c_S, \Delta, k)$ to the client C.

After receiving $(s_S, \textbf{\textit{hp}}_S, c_S, \Delta, k)$ from the server S, C computes $P_C \leftarrow \mathsf{PPrehash}(\textbf{\textit{param}}, \pi_C, s_P)$, $H_C \leftarrow \mathsf{PHash}(\textbf{\textit{param}}, s_S, P_C)$, $tk_{CS} = \mathsf{Hash}(\textbf{\textit{hk}}_C, L_{H_C}, c_S) \oplus \mathsf{ProjHash}(\textbf{\textit{hp}}_S, L_{(s_H, H_C)}, c_C, \pi_C, r_C)$, and checks if $k = H_1(id, S, \Delta, tk_{CS})$, if not, C rejects and aborts. Otherwise, C computes $k' = H_2(id, S, \Delta, tk_{CS})$, and decodes to obtain $sk = \mathsf{ECC}^{-1}(tk_{CS} \oplus \Delta)$. If $sk = \bot$, C rejects and aborts. Otherwise, C accepts $sk \leftarrow_r \{0, 1\}^{\kappa}$ as the shared session key. Then C sends k' to the server. The server receives k', if $k' = H_2(id, S, \Delta, tk_{SC})$, we know the client obtain matching session key, output sk.

In the second round of the protocol, the server S sends to the client C value k, which contains identities of the client and the server, the APSH value tk_{sc} and associated value Δ. If k equals to $H_1(id, S, \Delta, tk_{CS})$, then the client can consider the server as honest. In the third round of the protocol, the client sends to the server message k', which is generated by using tk_{CS}. Through this process, the client obtains matching session key. In this way, our protocol realizes mutual explicit authentication.

Correctness: We argue that honestly users can obtain the same session key with all but negligible probability. According to the ϵ-approximate correctness of ASPH, we know $\mathsf{Hash}(\boldsymbol{hk}_C, L_{H_C}, c_S) \oplus \mathsf{ProjHash}$ $(\boldsymbol{hp}_C, L_{H_S}, c_S, r_S) \in \{0, 1\}^{\ell}$ has at most ϵ-fraction non-zeros. Thus, we have $\mathsf{Ham}(tk_{SC}, tk_{CS}) \leq 2\epsilon$ with all but negligible probability. Since ECC can correct 2ϵ-fraction of errors by assumption, we have that $\mathrm{sk} = \mathsf{ECC}^{-1}$ $(tk_{CS} \oplus tk_{SC} \oplus \mathsf{ECC}(sk))$ holds.

6. Security model and security proof of our VPAKE protocol

In this section, we present a security model based on Ref. [27], then we examine the security of our protocol with this model.

6.1 Security model

Formally, the protocol relies on a setup assumption that a CRS and other public parameters are established by a trusted third party before any execution of the protocol.

6.1.1 Users and passwords

Let $U \in \mathcal{U}$ be the set of protocol users. For every distinct $C, S \in \mathcal{U}$, each client C holds a password pw_C, which is mapped to π_C, while each server S holds a random salt $s_H \leftarrow \boldsymbol{Salt}(pp)$ and a hash value $h = \boldsymbol{Hash}(pp, P, s_P, s_H)$ of π_C. We assume that each pw_C is chosen independently and uniformly from some dictionary set $\mathcal{D}$.

6.1.2 Protocol execution

The adversary $\mathcal{A}$ can make to any given instances $\prod_U^i$ during the game are as follows:

- **_Execute_**(C, i, S, j): This query models a passive attack in which the adversary eavesdrops an honest protocol execution between instances $\prod_C^i$ and $\prod_S^j$, and returns the transcript to $\mathcal{A}$.
- **_Send_**(U, i, m): This query models a positive attack in which the adversary intercept a message and modify it that be sent to instance $\prod_U^i$. We assume that $\mathcal{A}$ sees if the query causes $\prod_U^i$ to accept or terminate.

- **Reveal**(U, i): This query returns the session key sk_U^i to the adversary if it has been generated.
- **Corrupt**(S): This query models server corruption, and its output is the salt s_H and the hash value h. Any client C with password π_C is said to have his password hash corrupted.
- **Test**(U, i): This query chooses a bit $b \leftarrow_r \{0, 1\}$. If b = 1, it returns the session key sk_U^i of instance $\prod_U^i$ to $\mathcal{A}$; if b = 0, it returns a key chosen uniformly at random. The adversary is only allowed to query this oracle once.

6.1.3 Partnering and freshness

Instances $\prod_C^i$ and $\prod_S^j$ are partnered if $sid_C^i = sid_S^j \neq \perp$, $pid_C^i = S$ and $pid_S^i = C$. We say that a VPAKE protocol is correct if instances $\prod_C^i$ and $\prod_S^j$ are partnered, then we have that $acc_C^i = acc_S^j = 1$ and $sid_C^i = sid_S^j \neq \perp$ hold (sid_U^i, pid_U^i, sk_U^i denote the session id, partner id, and session key for instance $\prod_U^i$. acc_U^i and $term_U^i$ are Boolean variables denoting whether instance $\prod_U^i$ has accepted or terminated). We say that an instance $\prod_U^i$ is fresh if the conditions hold: (1) the adversary $\mathcal{A}$ did not make a Reveal(C, i) query to instance $\prod_C^i$; (2) the adversary $\mathcal{A}$ did not make a Reveal(S, j) query to instance $\prod_S^i$, where instances $\prod_C^i$ and $\prod_S^j$ are partnered; (3) the adversary $\mathcal{A}$ did not make a Corrupt(S) query before the Test query and did not make a **Send**(U, i, m) query for some m.

6.1.4 Definition of security

We say that the adversary $\mathcal{A}$ succeeds if $\mathcal{A}$ makes a Test query to a fresh instance $\prod_U^i$ only once, outputs a bit b', and $b' = b$ where b is the bit selected by the Test oracle. $\mathcal{A}$'s advantage is defined as:

$$Adv_{\Pi, \mathcal{A}}(\kappa) = 2\Pr[Succ] - 1 \tag{7}$$

A VPAKE protocol Π is secure if for all dictionaries $\mathcal{D}$ and for all adversaries $\mathcal{A}$ making at most $Q(\kappa)$ on-line attacks, it holds that $Adv_{\Pi, \mathcal{A}}(\kappa) \leq Q(\kappa)/|\mathcal{D}| + negl(\kappa)$.

6.2 Security proof

In this section, the security proof of the protocol is given according to the security model described in Section 5.

Theorem 2. *If* PKE $= ($KeyGen, Enc, Dec$)$ is a CCA-secure PKE scheme associated with an ϵ-approximate smooth projective hash $(K, \ell, \{$Hash$_{hk}:$ $X \rightarrow \{0, 1\}^{\ell}\}_{hk \in K}, HP,$ Proj:$K \rightarrow HP)$, *and* $\mathcal{H} = ($Setup, PPreSalt,

PPreHash, PSalt, PHash) *is a randomized password hashing scheme, ECC:* $\{0, 1\}^{\ell} \rightarrow \{0, 1\}^{\kappa}$ *is an error-correcting code which can correct* 2ϵ-*fraction of errors, then the above protocol is a secure VPAKE.*

Proof

Let Π denote the protocol in Fig. 2. We assume that a simulator controls all oracles to which the adversary has access to. The simulator runs protocol Π, including selecting passwords for each user and deriving verifier from the passwords. The adversary succeeds if it can guess the bit b chosen by the simulator during the Test query. We construct a sequence of games from G_0 to G_8, where G_0 is the real security game. The security is established by showing that the adversary's advantage in game G_0 and G_8 will differ at most $Q(\kappa)/|\mathcal{D}| +$ negl(κ). Let $\boldsymbol{Adv}_{\mathcal{A},i}(\kappa)$ be the adversary $\mathcal{A}$'s advantage in game G_i.

We divide the adversary's Send query into three types:

- $\boldsymbol{Send}_0(S^i, C^j, \boldsymbol{Start})$ query which enables an adversary to ask C^j to initiate an execution with S^i and return the first flow sent by C^j to S^i;
- $\boldsymbol{Send}_1(C^j, S^i, m)$ query which enables an adversary to send the first flow to S^j. It returns the second flow answered back by S^j;
- $\boldsymbol{Send}_2(S^i, C^j, m)$ query which enables an adversary to send the second flow to S^i and return nothing but define the session key sk.

Game G_1: We modify the executing of the password hashing scheme and the simulator stores the pre-salt s_P along with H and s_H. Under salt indistinguishability, we have $|\boldsymbol{Adv}_{\mathcal{A},1}(\kappa) - \boldsymbol{Adv}_{\mathcal{A},0}(\kappa)| = 0$.

Game G_2: We modify the Execute query that the values tk_{CS} is directly compute using the corresponding hash keys $\boldsymbol{hk}_C$ and $\boldsymbol{hk}_S$, i.e., $tk_{CS} = \mathsf{Hash}(\boldsymbol{hk}_C, L_{H_S}, c_S) \oplus \mathsf{Hash}(\boldsymbol{hk}_S, L_{(s_H, H_S)}, c_S)$.

Let $\left(K, \ell, \left\{\mathsf{Hash}_{hk} : X \rightarrow \{0, 1\}^{\ell}\right\}_{hk \in K}, HP, \mathsf{Proj} : K \rightarrow HP\right)$ be the ϵ-approximate SPH, and ECC: $\{0, 1\}^{\ell} \rightarrow \{0, 1\}^{\kappa}$ is an error-correcting code which can correct 2ϵ-fraction of errors, then we have $|\boldsymbol{Adv}_{\mathcal{A},2}(\kappa) - \boldsymbol{Adv}_{\mathcal{A},1}(\kappa)| \leq \boldsymbol{negl}(\kappa)$.

Game G_3: We modify the Execute query that the ciphertext c_S and c_C are replaced by the dummy password 0.

Since the encryption scheme $\mathsf{PKE} = (\mathsf{KeyGen}, \mathsf{Enc}, \mathsf{Dec})$ is a CCA-secure scheme based on lattice, we have $|\boldsymbol{Adv}_{\mathcal{A},3}(\kappa) - \boldsymbol{Adv}_{\mathcal{A},2}(\kappa)| \leq \boldsymbol{negl}(\kappa)$.

Game G_4: We modify the Execute query that the random coins of the session key are replaced.

Since the c_S and c_C do not fit the conditions of language, and the ASPH function is smooth, we have that $|\boldsymbol{Adv}_{\mathcal{A},4}(\kappa) - \boldsymbol{Adv}_{\mathcal{A},3}(\kappa)| \leq \boldsymbol{negl}(\kappa)$.

Game G_5: We modify the **Send**$_1$ query by using a decryption oracle, or alternatively knowing the decryption key. The message $m = (\mathbf{hp}_C, c_C)$ is sent, if the password hash value and salt s_P, s_H of S is corrupted, we answer and compute the session key. Otherwise, there are three cases:

- The message has been generated, we first decrypt the ciphertext to get the pre-hash value P used be the adversary:
  - If $\mathsf{Check}(pp, s_P, s_H, P, H_S) = 1$ with s_P, s_H, and H_S of server, then we choose the session key $sk = \perp$; if later the adversary ask the session key via a Test-query, the simulator succeeds and terminates the experiment;
  - Otherwise, the simulator chooses the session key sk at random;
- If the message m is a reply of a previously used message, the simulator knows the hashing key $\mathbf{hk}_C$ instead of $\mathbf{hp}_C$. Thus, the simulator can compute the session key sk using the hashing key $\mathbf{hk}_C$ rather than using $\mathbf{hp}_C$ and the randomness used to generate c_S.

The change in the first case can only increase the advantage of the adversary. While since the smoothness of our ASPH, the change in the second case is indistinguishable. And the change in the third case does not affect the advantage. Therefore, we have $\mathbf{Adv}_{A,4}(\kappa) \leq \mathbf{Adv}_{A,5}(\kappa) + \mathbf{negl}(\kappa)$.

Game G_6: We modify again the way **Send**$_1$ query is answered, the message $m = (\mathbf{hp}_C, c_C)$ is a reply of a previously used message, in response to the query **Send**$_1$ there are two cases:

- If instances S^j and C^i are partnered, then set the session key identical to the session key for S^j;
- Otherwise, chooses the session key sk uniformly.

The change in in the first case does not affect the advantage of the adversary. The change in the second case is indistinguishable, due to the smoothness of our ASPH, we have $|\mathbf{Adv}_{A,6}(\kappa) - \mathbf{Adv}_{A,5}(\kappa)| \leq \mathbf{negl}(\kappa)$.

Game G_7: We modify the **Send**$_2$ query. When a message $m = (s_H, \mathbf{hp}_S, c_S, \Delta, k)$ is sent to some client C^i by some server S^i, if the password π_C of client C has been corrupted, we answer honestly and compute the session key honestly. Otherwise, there are four cases:

- If the message m is not generated by S, after receiving the first message $(\mathbf{hp}_C, c_C)$ sent by C^i, then we first decrypt the ciphertext to get the hash value H used be the adversary:
  - If $H = \mathsf{PHash}(pp, s_P, s_H, \pi_C)$, then we choose the session key $sk = \perp$; if later the adversary ask the session key via a Test-query, we stop the simulation and let it win;
  - Otherwise we choose the session key sk at random;

- If the message m is generated by some instance S^i:
  - If S and C are partnered, then set the session key equal to the one computed by S^i;
  - Otherwise we choose the session key sk at random.

The change in the first case can only increase the advantage of the adversary, while the changes in the second and fourth cases are indistinguishable under the smoothness of the ASPH and thus only increase the advantage of the adversary by a negligible term. The change in the third case does not change the advantage of the adversary. Therefore, we have $Adv_{A,6}(\kappa) \leq Adv_{A,7}(\kappa) + negl(\kappa)$.

Game G_8: We compute c_C as encryption of the dummy password 0. c_C is generated as response to $Send_0$ query.

Since the encryption scheme PKE = (KeyGen, Enc, Dec) is a CCA-secure scheme based on lattice, we have $|Adv_{A,8}(\kappa) - Adv_{A,7}(\kappa)| \leq negl(\kappa)$.

In this last game, the simulator does not use the password. The view of A is independent of the user's password. A succeeds only when it can generate message that corresponds to an encryption of the correct password. For any adversary A that makes at most $Q(\kappa)$ times on–line attacks, and from game G_0 to G_8, we have that $Adv_{A,0}(\kappa) \leq Q(\kappa)/|\mathcal{D}| + negl(\kappa)$. This completes the proof of Theorem 2.

7. Proof of Theorem 3

We restate Theorem 3 as it appears in Ref. [54] and then provide proof of it as below.

ZKAoK protocol for the relation $R_{abstract}$:

1. Commitment: Prover samples $\mathbf{r}_\omega \overset{\$}{\leftarrow} \mathbb{Z}_q^\ell$, $\phi \overset{\$}{\leftarrow} S$ and randomness ρ_1, ρ_2, ρ_3 for COM.

   Then, a commitment $CMT = (C_1, C_2, C_3)$ is sent to the verifier $\mathcal{V}$, where
   $$C_1 = COM(\phi, \mathbf{M} \cdot \mathbf{r}_\omega \bmod q; \rho_1), \quad C_2 = COM(\Gamma_\phi(\mathbf{r}_\omega); \rho_2), \quad C_3 = COM(\Gamma_\phi(\mathbf{w} + \mathbf{r}_\omega \bmod q); \rho_3).$$

2. Challenge: $\mathcal{V}$ sends a challenge $Ch \overset{\$}{\leftarrow} \{1, 2, 3\}$ to prover.
3. Response: Based on Ch, prover sends RSP computed as follows:
   - $Ch = 1$: Let $\mathbf{t}_\omega = \Gamma_\phi(\mathbf{w})$, $\mathbf{t}_r = \Gamma_\phi(\mathbf{r})$, and $RSP = (\mathbf{t}, \mathbf{t}_r, \rho_2, \rho_3)$.
   - $Ch = 2$: Let $\phi_2 = \phi$, $\mathbf{w}_2 = \mathbf{w} + \mathbf{r} \bmod q$, and $RSP = (\phi_2, \mathbf{w}_2, \rho_1, \rho_3)$.
   - $Ch = 2$: Let $\phi_3 = \phi$, $\mathbf{w}_3 = \mathbf{r}$, and $RSP = (\phi_3, \mathbf{w}_3, \rho_1, \rho_2)$.

Verification: Receiving RSP, $\mathcal{V}$ proceeds as follows:

- $Ch=1$: Check that $\mathbf{t}_\omega \in \boldsymbol{VALID}$, $C_2=\text{COM}(\mathbf{t}_r; \rho_2)$, $C_3=\text{COM}(\mathbf{t}_\omega + \mathbf{t}_r \bmod q); \rho_3)$.
- $Ch=2$: Check that $C_1=\text{COM}(\phi_2, \mathbf{M}\cdot\mathbf{w}_2 - \mathbf{v} \bmod q; \rho_1)$, $C_3=\text{COM}(\Gamma_{\phi 2}(\mathbf{w}_2); \rho_3)$.
- $Ch=3$: Check that $C_1=\text{COM}(\phi_2, \mathbf{M}\cdot\mathbf{w}_3; \rho_1)$, $C_2=\text{COM}(\Gamma_{\phi 3}(\mathbf{w}_3); \rho_2)$.

In each case, $\mathcal{V}$ outputs 1 if and only if all the conditions hold.

Theorem 3. *The protocol is a statistical zero-knowledge argument of knowledge (ZKAoK) with perfect completeness, soundness error 2/3, and communication cost* $\mathcal{O}(\ell \log q)$. *Namely:*

- There exists a polynomial-time simulator that, on input $(\mathbf{M}, \mathbf{v})$, outputs an accepted transcript statistically close to that produced by the real prover.
- There exists a polynomial-time knowledge extractor that, on input a commitment CMT and three valid responses $(\text{RSP}_1, \text{RSP}_2, \text{RSP}_3)$ to all three possible values of the challenge Ch, outputs $\mathbf{w}' \in \text{VALID}$ such that $\mathbf{M}\cdot\mathbf{w}' = \mathbf{v} \bmod q$.

Proof. Perfect completeness of the protocol can be checked: If a prover follows the protocol honestly, the verifier will always accept. It is also easy to see that the communication cost is bounded by $(\ell \log q)$.

We now prove that the protocol is a statistical zero-knowledge argument of knowledge.

Zero-Knowledge Property. We construct a PPT simulator SIM interacting with a (possibly dishonest) verifier $\widehat{\mathcal{V}}$, such that, given only the public inputs, with probability negligibly close to 2/3, SIM outputs a simulated transcript that is statistically close to the one produced by the honest prover in the real interaction. SIM first chooses uniformly at random, $\overline{Ch} \in \{1, 2, 3\}$, its prediction of Ch that $\widehat{\mathcal{V}}$ will not choose.

Case $Ch=1$: Using basic linear algebra over $\mathbb{Z}_q$, SIM computes a vector $\mathbf{w}' \in \mathbb{Z}_q^\ell$ such that $\mathbf{M}\cdot\mathbf{w}' = \mathbf{v} \bmod q$. Next, it samples $\mathbf{r}_\omega \xleftarrow{\$} \mathbb{Z}_q^\ell$, $\phi \xleftarrow{\$} S$ and randomness ρ_1, ρ_2, ρ_3 for COM. Finally, it sends the commitment CMT $= (C_1', C_2', C_3')$ to $\widehat{\mathcal{V}}$, where.

$$C_1' = \text{COM}(\phi, \mathbf{M}\cdot\mathbf{r}_\omega; \rho_1), \quad C_2' = \text{COM}(\Gamma_\phi(\mathbf{r}_\omega); \rho_2), \quad C_3' = \text{COM}(\Gamma_\phi(\mathbf{w}' + \mathbf{r}_\omega); \rho_3).$$

Receiving a challenge Ch from $\widehat{\mathcal{V}}$, the simulator responds as follows:

- If $Ch=1$: Output $\perp$ and abort.
- If $Ch=2$: Send RSP $= (\phi, \mathbf{w}' + \mathbf{r}_\omega, \rho_1, \rho_3)$.
- If $Ch=3$: Send RSP $= (\phi, \mathbf{r}_\omega, \rho_1, \rho_2)$.

Case $\overline{Ch} = 2$: SIM samples $\mathbf{w}' \xleftarrow{\$} \text{VALID}$, $\mathbf{r}_\omega \xleftarrow{\$} \mathbb{Z}_{q'}^\ell$, $\phi \xleftarrow{\$} S$ and randomness ρ_1, ρ_2, ρ_3 for COM. Then, it sends the commitment $\text{CMT} = (C_1', C_2', C_3')$ to $\widehat{\mathcal{V}}$, where $C_1' = \text{COM}(\phi, \mathbf{M} \cdot \mathbf{r}_\omega; \rho_1)$, $C_2' = \text{COM}(\Gamma_\phi(\mathbf{r}_\omega); \rho_2)$, $C_3' = \text{COM}(\Gamma_\phi(\mathbf{w}' + \mathbf{r}_\omega); \rho_3)$.

Receiving a challenge Ch from $\widehat{\mathcal{V}}$, the simulator responds as follows:

- If $Ch = 1$: Send $\text{RSP} = (\Gamma_\phi(\mathbf{w}'), \Gamma_\phi(\mathbf{r}_\omega), \rho_2, \rho_3)$ Output $\perp$ and abort.
- If $Ch = 2$: Output $\perp$ and abort.
- If $Ch = 3$: Send $\text{RSP} = (\phi, \mathbf{r}_\omega, \rho_1, \rho_2)$.

Case $\overline{Ch} = 3$: SIM samples $\mathbf{w}' \xleftarrow{\$} \text{VALID}$, $\mathbf{r}_\omega \xleftarrow{\$} \mathbb{Z}_{q'}^\ell$, $\phi \xleftarrow{\$} S$ and randomness ρ_1, ρ_2, ρ_3 for COM. Then, it sends the commitment $\text{CMT} = (C_1', C_2', C_3')$ to $\widehat{\mathcal{V}}$, where $C_2' = \text{COM}(\Gamma_\phi(\mathbf{r}_\omega); \rho_2)$, $C_3' = \text{COM}(\Gamma_\phi(\mathbf{w}' + \mathbf{r}_\omega); \rho_3)$ as in the previous two cases, while $C_1' = \text{COM}(\phi, \mathbf{M} \cdot (\mathbf{w}' + \mathbf{r}_\omega) - \mathbf{v}; \rho_1)$.

Receiving a challenge Ch from $\widehat{\mathcal{V}}$, it responds as follows:

- If $Ch = 1$: Send RSP computed as in the case $(\overline{Ch} = 2, \overline{Ch} = 1)$.
- If $Ch = 2$: Send RSP computed as in the case $(\overline{Ch} = 1, \overline{Ch} = 2)$.
- If $Ch = 3$: Output $\perp$ and abort.

In each of the cases above, since COM is statistically hiding, the distribution of the commitment CMT and challenge Ch from $\widehat{\mathcal{V}}$ are statistically close to those in the real interaction.

Hence, the probability that the simulator outputs $\perp$ is negligibly close to $1/3$. Moreover, whenever the simulator does not halt, it will provide an accepted transcript, the distribution of which is statistically close to the prover's in the real interaction. In other words, the constructed simulator can successfully impersonate the honest prover with probability negligibly close to $2/3$.

Suppose $\text{RSP}_1 = (\mathbf{t}_\omega, \mathbf{t}_r, \rho_2, \rho_3)$, $\text{RSP}_2 = (\phi_2, \mathbf{w}_2, \rho_1, \rho_3)$ and $\text{RSP}_3 = (\phi_3, \mathbf{w}_3, \rho_1, \rho_2)$ are three valid responses to the same commitment $\text{CMT} = (C_1, C_2, C_3)$, with respect to all three possible values of the challenge. The validity of these responses implies that:

$$
\begin{cases}
\mathbf{t}_\omega \in \text{VALID}; \\
C_1 = \text{COM}(\phi_2, \mathbf{M} \cdot \mathbf{w}_2 - \mathbf{v} \bmod \mathrm{q}; \rho_1) = \text{COM}(\phi_3, \mathbf{M} \cdot \mathbf{w}_3; \rho_1); \\
C_2 = \text{COM}(\mathbf{t}_r; \rho_2) = \text{COM}(\Gamma_{\phi_3}(\mathbf{w}_3); \rho_2); \\
C_3 = \text{COM}(\mathbf{t}_\omega + \mathbf{t}_r \bmod q; \rho_3) = \text{COM}(\Gamma_{\phi_2}(\mathbf{w}_2); \rho_3).
\end{cases}
$$

$$(8)$$

Since COM is computationally binding, it implies that

$$
\begin{cases}
\mathbf{t}_\omega \in \text{VALID}; \phi_2 = \phi_3; \mathbf{t}_r = \Gamma_{\phi 3}(\mathbf{w}_3); \mathbf{t}_\omega + \mathbf{t}_r = \Gamma_{\phi 2}(\mathbf{w}_2) \bmod q; \\
\mathbf{M} \cdot \mathbf{w}_2 - \mathbf{v} = \mathbf{M} \cdot \mathbf{w}_3 \bmod q).
\end{cases}
\tag{9}
$$

Since $\mathbf{t}_\omega \in \text{VALID}$, if we let $\mathbf{w}' = [\Gamma_{\phi 2}]^{-1}(\mathbf{t}_\omega)$, then $\mathbf{w}' \in \text{VALID}$. Furthermore, we have $\Gamma_{\phi 2}(\mathbf{w}') + \Gamma_{\phi 2}(\mathbf{w}_3) = \Gamma_{\phi 2}(\mathbf{w}_2) \bmod q$, which means that $\mathbf{w}' + \mathbf{w}_3 = \mathbf{w}_2 \bmod q$, and $\mathbf{M} \cdot \mathbf{w}' + \mathbf{M} \cdot \mathbf{w}_3 = \mathbf{M} \cdot \mathbf{w}_2 \bmod q$. As a result, we have $\mathbf{M} \cdot \mathbf{w}' = \mathbf{v} \bmod q$, concluding the proof.

8. Summary of key lessons learned and research directions in the field

Based on our construction of the new VPAKE protocol, we learned the following lessons:

- In order to prove the security of the protocol, various aspects of the protocol need to be taken into consideration, and different game needs to be established according to different scenario.
- In the process of protocol constructing, the efficiency level should also be considered while security level is pursued. The selection of parameters have great impact on the calculation and communication efficiency of the protocol.
- The approximate smoothed projection function in the protocol should be designed according to the underlying PKE. And this function can be instantiated on lattice.

There are many research directions in PAKE. One is research on PAKE based on smooth projective hash. Our study fall into this group: we combine public-key encryption scheme based on LWE with the associated approximate smooth projective hash to construct a PAKE based on lattices. The PKEs with associated SPH (based on either Decision Diffie–Hellman (DDH) or decisional linear assumptions) can be used to instantiate generic framework. Another direction is research on PAKE based on Key Encapsulation Mechanism (KEM). Asymmetric key encapsulation (also known as key transport) can be used in this group of approaches. With asymmetric key encapsulation, the sender transmit a random cryptographic key K using the receiver's public key, and K can only be recovered only by the intended receiver. Hybrid encryption, by which the participants use symmetric algorithms to encrypt and/or authenticate big data under K, is employed in this group of approaches. In future, more exploration can be conducted in these two directions.

9. Conclusion

PAKE protocols have advantages from both theoretical and practical perspectives. They are unique, as they allow the secure computation and authentication of a high-entropy piece of data using a low-entropy string as a starting point. The server stores a verifier, which is the hash value of passwords with a salt, so leakage of information from server side can be overcome. From practical application point of view, in PAKE protocols, most passwords have low entropy, so these passwords can be remembered easily. This advantage is particularly appealing in the condition where most people are accessing sensitive personal data remotely from their handheld devices. Thus, PAKEs have great potential of being put into practice.

We propose a VPAKE protocol based on the LWE. It is the first VPAKE based on lattices. Our new protocol employs ZKPPC to process privacy-preserving password-based authentication. This technique is to create a password encoding mechanism, which can and interacts smoothly with the lattice-based cryptographic tools. Furthermore, we construct an associated ASPH function, which suits our VPAKE protocol. We provide proof of security of our PAKE protocol with the Random Oracle Model (ROM). Since this security proving approach does not apply in a quantum adversary context, we need to further test the security level of our protocol with the Quantum Random Oracle Model (QROM) in our future research.

Acknowledgment

This work is supported by "The 13th Five-Year Plan" National Crypto Development Fund (No. MMJJ20170122), the National Natural Science Foundation of China (No. 61802117), Basic and Frontier Technology Research Projects of Henan Science and Technology Department (No. 142300410147, 182102310923), Natural Science Foundation of Henan Polytechnic University (No. T2018-1).

Key terminology and definitions

Discrete Logarithm Problem Let G be any group. Denote its group operation by multiplication and its identity element by 1. Let b be any element of G. For any positive integer k, the expression b^k denotes the product of b with itself k times: $b^k = \underbrace{b \cdot b \cdot \ldots \cdot b}_{k \text{ times}}$.
Similarly, let b^{-k} denote the product of b^{-1} with itself k times. For $k=0$, the kth power is the identity: $b^0 = 1$. Let a also be an element of G. An integer k that solves the equation $b^k = a$ is termed a discrete logarithm (or simply logarithm, in this context) of a to the base b. One writes $k = \log_b a$.

Chosen–Ciphertext Attack (CCA) A CCA is an attack model for cryptanalysis where the cryptanalyst can gather information by obtaining the decryptions of chosen ciphertext.

By these pieces of information, the adversary can attempt to recover the hidden secret key used for decryption.

RSA It was developed by Ron Rivest, Adi Shamirh, and Len Adleman at Massachusetts Institute of Technology. RSA is an algorithm for public-key encryption. It is the first algorithm known to be suitable for encryption and digital signing.

ElGamal Encryption In cryptography, the ElGamal encryption system is an asymmetric key encryption algorithm for public-key cryptography. It is based on the Diffie-Hellman key exchange. The system provides an additional layer of security by asymmetrically encrypting keys, which are previously used for symmetric message encryption. ElGamal encryption is used in the free GNU Privacy Guard software, recent versions of PGP, and other cryptosystems.

Elliptic-Curve Cryptography (ECC) ECC is a public-key cryptography method based on the algebraic structure of elliptic curves over finite fields. ECC requires smaller keys compared to non-EC cryptography (based on plain Galois fields). Elliptic curves are applicable for key agreement, digital signatures, pseudo-random generators and other tasks.

Bilinear Pairing The bilinear pairing, namely the weil-pairing and the tate-pairing of algebraic curves, are important tools for research on algebraic geometry.

Brute-Force Attack In cryptanalysis and computer security, password cracking is the process of recovering passwords from data that have been stored in or transmitted by a computer system. A common approach (brute-force attack) is to guess repeatedly the password and check them against an available cryptographic hash of the password.

Parity Check Parity Check is a method of verifying the correctness of a code transfer. The check is carried out according to whether the number of "1" in the transmitted binary code is odd or even. An odd number is called an odd check, whereas an even check is called an even check.

Gram-Schmidt Orthogonalization It is a process for constructing an orthogonal basis for a Euclidean space, given any basis for the space.

Coding Theory It is study of the properties of codes and their respective fitness for specific applications. Codes are used for data compression, cryptography, error detection and correction, data transmission and data storage. Study of codes are conducted in various scientific disciplines, including information theory, electrical engineering, and mathematics, linguistics, and computer science. The purpose of study on codes is to design efficient and reliable data transmission methods.

Chernoff Bound In probability theory, the Chernoff bound gives exponentially decreasing bounds on tail distributions of sums of independent random variables. It is a sharper bound than the known first- or second-moment-based tail bounds such as Markov's inequality or Chebyshev's inequality, which only yield power-law bounds on tail decay.

Decision Diffie-Hellman (DDH) The decision Diffie-Hellman problem is a core computational problem in cryptography. It is known that the Weil and Tate pairings can be used to solve many DDH problems on elliptic curves. DDH enables one to construct cryptographic systems with strong security properties.

Key Encapsulation Mechanism (KEM) KEM is a group of encryption techniques, which is designed to secure symmetric cryptographic key material during transmission. KEM uses asymmetric (public-key) algorithms.

Hybrid Cryptosystem In cryptography, a hybrid cryptosystem combines the convenience of a public-key cryptosystem with the efficiency of a symmetric-key cryptosystem. Public-key cryptosystems are convenient, as they do not require the sender and the receiver to share a common secret in order to communicate securely. However, they often rely on complicated mathematical computations and are thus much more inefficient than comparable symmetric-key cryptosystems. In many applications, a public-key cryptosystem has a high cost on encrypting long messages. This problem can be resolved by employing hybrid systems.

Random Oracle In cryptography, a random oracle is an oracle (a theoretical black box) that responds to every unique query with a (truly) random response chosen uniformly from its output domain. If a query is repeated, it responds the same way every time that query is submitted.

References

[1] S.M. Bellovin, M. Merritt, Encrypted key exchange: password-based protocols secure against dictionary attacks, in: Proceedings 1992 IEEE Computer Society Symposium on Re-search in Security and Privacy, IEEE, Oakland, 1992, pp. 72–84.

[2] P.W. Shor, Algorithms for quantum computation: discrete logarithms and factoring, in: Proceeding of the 35th Symposium on the Foundation of Computer Science, IEEE, Santa Fe, 1994, pp. 124–134.

[3] M. Ajtai, Generating hard instances of lattice problems (extended abstract), in: STOC 1996: Proceedings of the Twenty-Eighth Annual ACM Symposium on Theory of Computing, Association for Computing Machinery, New York, 1996, pp. 99–108.

[4] D. Micciancio, Generalized compact knapsacks, cyclic lattices, and efficient one-way functions from worst-case complexity assumptions, in: Proceedings of the 43rd Symposium on Foundations of Computer Science, IEEE, Vancouver, 2002, pp. 356–365.

[5] C. Peikert, A. Rosen, Efficient collision-resistant hashing from worst-case assumptions on cyclic lattices, in: S. Halevi, T. Rabin (Eds.), TCC 2006, Lecture Notes in Computer Science, Springer, Heidelberg, 2006, pp. 145–166.

[6] D. Boneh, D.M. Freeman, Linearly homomorphic signatures over binary fields and new tools for lattice-based signatures, in: D. Catalano, N. Fazio, R. Gennaro, A. Nicolosi (Eds.), Public Key Cryptography, Lecture Notes in Computer Science, Springer, Heidelberg, 2006, pp. 1–16.

[7] X. Boyen, Lattice mixing and vanishing trapdoors: a framework for fully secure short signatures and more, in: P.Q. Nguyen, D. Pointcheval (Eds.), Public Key Cryptography, Lecture Notes in Computer Science, Springer, Heidelberg, 2010, pp. 499–517.

[8] R. Canetti, S. Halevi, J. Katz, Y. Lindell, P. MacKenzie, Universally composable password-based key exchange, in: R. Cramer (Ed.), EUROCRYPT 2005, Lecture Notes in Computer Science, Springer, Heidelberg, 2005, pp. 404–421.

[9] M. Ajtai, C. Dwork, A public-key cryptosystem with worst-case/average-case equivalence, in: STOC97: 29th Annual Symposium on Theory of Computing, Association for Computing Machinery, New York, 1997, pp. 284–293.

[10] O. Regev, New lattice-based cryptographic constructions, J. ACM 51 (2004) 899–942.

[11] O. Regev, On lattices, learning with errors, random linear codes, and cryptography, J. ACM 56 (2009) 1–40.

[12] C. Gentry, C. Peikert, V. Vaikuntanathan, Trapdoors for hard lattices and new cryptographic constructions, in: STOC '08: 40th Annual Symposium on Theory of Computing, Association for Computing Machinery, New York, 2008, pp. 197–206.

[13] D. Cash, D. Hofheinz, E. Kiltz, C. Peikert, Bonsai trees, or how to delegate a lattice basis, in: H. Gilbert (Ed.), EUROCRYPT 2010, Lecture Notes in Computer Science, Springer, Heidelberg, 2010, pp. 523–552.

[14] S. Agrawal, D. Boneh, X. Boyen, Lattice basis delegation in fixed dimension and shorter-ciphertext hierarchical IBE, in: T. Rabin (Ed.), CRYPTO 2010, Lecture Notes in Computer Science, vol. 6223, Springer, Heidelberg, 2010, pp. 98–115.

[15] S. Agrawal, D. Boneh, X. Boyen, Efficient lattice (H)IBE in the standard model, in: H. Gilbert (Ed.), EUROCRYPT 2010, Lecture Notes in Computer Science, vol. 6110, Springer, Heidelberg, 2010, pp. 553–572.

[16] J. Katz, V. Vaikuntanathan, Smooth projective hashing and password-based authenticated key exchange from lattices, in: M. Matsui (Ed.), ASIACRYPT 2009, Lecture Notes in Computer Science, Springer, Heidelberg, 2009, pp. 636–652.

[17] C. Peikert, Lattice cryptography for the internet, in: M. Mosca (Ed.), Post-Quantum Cryptography. PQCrypto 2014, Lecture Notes in Computer Science, Springer, Cham, 2014, pp. 197–219.

[18] W. Wen, L. Wang, A strongly secure lattice-based key exchange protocol, J. Com. Res. Dev. 51 (2015) 2258–2269.

[19] S. Agrawal, X. Boyen, V. Vaikuntanathan, P. Voulgaris, H. Wee, Functional encryption for threshold functions (or fuzzy IBE) from lattices, in: M. Fischlin, J. Buchmann, M. Manulis (Eds.), Public Key Cryptography, Lecture Notes in Computer Science, Springer, Heidelberg, 2012, pp. 280–297.

[20] C. Gentry, Fully homomorphic encryption using ideal lattices, in: STOC'09: 41th Annual Symposium on Theory of Computing, Association for Computing Machinery, New York, 2009, pp. 169–178.

[21] C. Gentry, Toward basing fully homomorphic encryption on worst-case hardness, in: T. Rabin (Ed.), CRYPTO 2010, Lecture Notes in Computer Science, Springer, Heidelberg, 2010, pp. 116–137.

[22] Z. Brakerski, V. Vaikuntanathan, Efficient fully homomorphic encryption from (standard) LWE, in: 2011 IEEE 52nd Annual Symposium on Foundations of Computer Science, IEEE, Palm Springs, 2011, pp. 97–106.

[23] Z. Brakerski, V. Vaikuntanathan, Fully homomorphic encryption from ring-LWE and security for key dependent messages, in: P. Rogaway (Ed.), CRYPTO 2011, Lecture Notes in Computer Science, Springer, Heidelberg, 2011, pp. 505–524.

[24] W. Diffie, M. Hellman, New directions in cryptography, IEEE Trans. Inf. Theory 22 (1976) 644–654.

[25] L. Gong, T.M.A. Lomas, R.M. Needham, J.H. Saltzer, Protecting poorly chosen secrets from guessing attacks, IEEE J. Sel. Areas Commun. 11 (1993) 648–656.

[26] S. Halevi, H. Krawczyk, Public-key cryptography and password protocols, ACM Trans. Inf. Syst. Secur. 2 (1999) 230–268.

[27] M. Bellare, D. Pointcheval, P. Rogaway, Authenticated key exchange secure against dictionary attacks, in: B. Preneel (Ed.), Advances in Cryptology—EUROCRYPT 2000, Lecture Notes in Computer Science, Springer, Heidelberg, 2000, pp. 139–155.

[28] V. Boyko, P.D. MacKenzie, S. Patel, Provably secure password-authenticated key exchange using Diffie-Hellman, in: B. Preneel (Ed.), Advances in Cryptology—EUROCRYPT 2000, Lecture Notes in Computer Science, Springer, Heidelberg, 2000, pp. 156–171.

[29] R. Gennaro, Y. Lindell, A framework for password-based authenticated key exchange, ACM Trans. Inf. Syst. Secur. 9 (2006) 181–234.

[30] P. MacKenzie, S. Patel, R. Swaminathan, Password-authenticated key exchange based on RSA, in: T. Okamoto (Ed.), ASIACRYPT 2000, Lecture Notes in Computer Science, Springer, Heidelberg, 2000, pp. 599–613.

[31] J. Katz, V. Vaikuntanathan, Round-optimal password-based authenticated key exchange, in: Y. Ishai (Ed.), TCC 2011, Lecture Notes in Computer Science, Springer, Heidelberg, 2011, pp. 293–310.

[32] J. Zhang, Y. Yu, Two-round PAKE from approximate SPH and instantiations from lattices, in: T. Takagi, T. Peyrin (Eds.), Advances in Cryptology—ASIACRYPT 2017, Lecture Notes in Computer Science, Springer, Cham, 2017, pp. 37–67.

[33] O. Goldreich, Y. Lindell, Session-key generation using human passwords only, J. Cryptol. 19 (2006) 241–340.
[34] E. Bresson, O. Chevassut, D. Pointcheval, Security proofs for an efficient password-based key exchange, in: Proceedings of the 10th ACM Conference on Computer and Communications Security, ACM Press, Washington, 2003, pp. 27–30.
[35] J. Katz, R. Ostrovsky, M. Yung, Efficient and secure authenticated key exchange using weak passwords, J. ACM 57 (2009) 1–39.
[36] R. Gennaro, Y. Lindell, A framework for password-based authenticated key exchange, in: E. Biham (Ed.), EUROCRYPT 2003, Lecture Notes in Computer Science, Springer, Heidelberg, 2003, pp. 524–543.
[37] R. Cramer, V. Shoup, Universal hash proofs and a paradigm for adaptive chosen ciphertext secure public-key encryption, in: L.R. Knudsen (Ed.), EUROCRYPT 2002, Lecture Notes in Computer Science, Springer, Heidelberg, 2002, pp. 45–64.
[38] M. Abdalla, F. Benhamouda, D. Pointcheval, Disjunctions for hash proof systems: new constructions and applications, in: E. Oswald, M. Fischlin (Eds.), EUROCRYPT 2015, Lecture Notes in Computer Science, Springer, Heidelberg, 2015, pp. 69–100.
[39] M. Abdalla, C. Chevalier, D. Pointcheval, Smooth projective hashing for conditionally extractable commitments, in: S. Halevi (Ed.), CRYPTO 2009, Lecture Notes in Computer Science, Springer, Heidelberg, 2009, pp. 671–689.
[40] F. Benhamouda, O. Blazy, C. Chevalier, D. Pointcheval, D. Vergnaud, New techniques for SPHFs and efficient one-round PAKE protocols, in: R. Canetti, J.A. Garay (Eds.), CRYPTO 2013, Lecture Notes in Computer Science, Springer, Heidelberg, 2013, pp. 449–475.
[41] A. Groce, J. Katz, A new framework for efficient password-based authenticated key exchange, in: Proceedings of the 17th ACM Conference on Computer and Communications Security, Association for Computing Machinery, New York, 2010, pp. 516–525.
[42] S.M. Bellovin, M. Merritt, Augmented encrypted key exchange: a password-based protocol secure against dictionary attacks and password file compromise, in: CCS93: 1st ACM Conference on Computer and Communications Security, Association for Computing Machinery, New York, 1993, pp. 244–250.
[43] M. Abdalla, O. Chevassut, D. Pointcheval, One-time verifier-based encrypted key exchange, in: S. Vaudenay (Ed.), Public Key Cryptography—PKC 2005. PKC 2005, Lecture Notes in Computer Science, Springer, Heidelberg, 2005, pp. 47–64.
[44] S.W. Lee, H.S. Kim, K.Y. Yoo, Efficient verifier-based key agreement protocol for three parties without server's public key, Appl. Math. Comput. 167 (2005) 996–1003.
[45] J.O. Kwon, J.Y. Hwang, C.W. Kim, D.H. Lee, Cryptanalysis of Lee–Kim–Yoo password-based key agreement scheme, Appl. Math. Comput. 168 (2005) 858–865.
[46] J.O. Kwon, I.R. Jeong, K. Sakurai, Efficient verifier-based password-authenticated key exchange in the three-party setting, Comput. Stand. Interfaces 29 (2007) 513–520.
[47] Y. Liu, H. Peng, J. Wang, Verifiable diversity ranking search over encrypted outsourced data, Comput. Mater. Con. 55 (2018) 037–057.
[48] X. Yang, H. Jiang, Q. Xu, A provably-secure and efficient verifier-based anonymous password-authenticated key exchange protocol, in: 2016 IEEE Trustcom/BigDataSE/ISPA, IEEE, Tianjin, 2016, pp. 670–677.
[49] F. Benhamouda, D. Pointcheval, Verifier-based password authenticated key exchange: new models and constructions, in: Cryptology ePrint archive: report 2013/833, IACR, 2014.
[50] F. Kiefer, M. Manulis, Zero-knowledge password policy checks and verifier-based PAKE, in: M. Kutyłowski, J. Vaidya (Eds.), Computer Security—ESORICS 2014. ESORICS 2014, Lecture Notes in Computer Science, Springer, Cham, 2014, pp. 295–312.
[51] E.B. Barker, W.E. Burr, L. Chen, SP 800-132. Recommendation for Password-Based Key Derivation: Part 1: Storage Applications, National Institute of Standards & Technology, Gaithersburg, 2010.

[52] D. Micciancio, S. Goldwasser, Complexity of lattice problems: a cryptographic perspective, in: The Kluwer International Series in Engineering and Computer Science, Kluwer Academic Publishers, New York, 2002.

[53] K. Nguyen, B.H.M. Tan, H. Wang, Zero-knowledge password policy check from lattices, in: P. Nguyen, J. Zhou (Eds.), Information Security. ISC 2017, Lecture Notes in Computer Science, Springer, Cham, 2017, pp. 92–113.

[54] B. Libert, S. Ling, K. Nguyen, H. Wang, Zero-knowledge arguments for lattice-based accumulators, logarithmic-size ring signatures and group signatures without trapdoors, in: M. Fischlin, J.S. Coron (Eds.), Advances in Cryptology—EUROCRYPT 2016. EUROCRYPT 2016, Lecture Notes in Computer Science, Springer, Heidelberg, 2016, pp. 1–31.

About the authors

Dr. Jinxia Yu is a Professor in the College of Computer Science and Technology at Henan Polytechnic University. She won the Outstanding Young Science and Technology Expert Award at the city of Jiaozuo, China. She has been in charge or participated in more than 10 projects of the National Natural Science Foundation of China. Her research interest include AI, intelligent information processing, and information security.

Miss. Huanhuan Lian is a Master student in the College of Computer Science and Technology in Henan Polytechnic University. Her research interests include Cryptography, information security and lattice cryptography. She won the National Scholarship and the first-class academic Scholarship of Henan Polytechnic University.

Dr. Zongqu Zhao obtained his Ph.D. degree from School of Computer Science and Technology, Si Chuan University, China, in 2015. Now he is a lecturer at School of Computer Science and Technology, Henan Polytechnic University. His research interests include cryptology, security of computer networks and malware analysis. He has published more than 10 articles in academic journals and international conferences.

Dr. Xiaojun Wang obtained his BEng in Computer and Communications from Beijing University of Posts and Telecommunications (BUPT), China in 1984 and his MEng in Computer Applications in 1987 also from BUPT. Xiaojun was employed as Assistant Lecturer/Lecturer in BUPT from 1987 to 1989. He received a PhD scholarship from the Sino-British Technical Co-operation Training Award in 1989, and did his PhD research in the School of Engineering in Staffordshire University (Staffordshire Polytechnic), England, and UK from 1989 to 1992. He joined the School of Electronic Engineering of Dublin City University as an assistant Lecturer in November 1992, and he is now a Professor. He was Head of China Affairs in Dublin City University for 5 years between 2002 and 2007. He was the main sponsor and organizer of the China-Ireland International Conference on Information and Communication Technology (CIICT 2006–2014). His research focuses on the innovations of ICT, which includes energy efficient ICT, information and network security, deep packet inspection, FPGA hardware acceleration for network security applications, data mining, energy-efficient network processor design, Internet of things, HDL modeling for synthesis and rapid FPGA prototyping, quantum secure communication.

Dr. Yongli Tang is Professor in the College of Computer Science and Technology at Henan Polytechnic University. He won the Outstanding Young Science and Technology Expert Award at the city of Jiaozuo. He is a member of the Education and Science Working Committee of Cryptography Society, the Henan comprehensive evaluation expert. His research areas include information security and cryptography algorithms. He has participated in the National 863 Program, and has participated or be responsible for more than 20 special projects for national information security strategy research and standard setting. He has published more than 30 papers in international journals and conferences. He also published two textbooks.

CHAPTER FIVE

Fingerprint liveness detection using an improved CNN with the spatial pyramid pooling structure

Chengsheng Yuan[a,b], Qi Cui[a], Xingming Sun[a], Q.M. Jonathan Wu[b], and Sheng Wu[c,*]

[a]School of Computer and Software, Nanjing University of Information Science and Technology, Nanjing, China
[b]Department of Electrical and Computer Engineering, University of Windsor, Windsor, ON, Canada
[c]Chinese Academy of Sciences, Beijing, China
[*]Corresponding author: e-mail address: wusheng12b@mails.ucas.edu.cn

Contents

Abstract

While fingerprint identification systems have been widely applied to daily life, how to protect them from presentation attacks has become a hot topic in the field of biometric verification. A feasible strategy of fingerprint recognition, called Fingerprint Liveness Detection (FLD), has attracted a lot of attention from researchers. Convolutional Neural Network (CNN) technology has been widely adopted by researchers in FLD, since it can automatically learn high-level semantic features from a large number of labeled samples. However, the main problem of CNN is that it only takes image of a fixed-size/ scale as input to the model. Therefore, two common techniques, crop and warp operation, have been devised. Due to the lack of some discriminative spatial information, the final detection performance of the model is not ideal. In this chapter, a novel FLD method based on an improved CNN with Spatial Pyramid Pooling (SPP) is proposed to overcome the limitation of image size/scale. Moreover, in order to resolve the problem of insufficient images, pre-train model parameters are adopted using the auxiliary ImageNet 2012 dataset, which serve as the initialization of our model parameter. Afterwards, we fine-tune our model parameters, using training samples from LivDet 2011 and LivDet 2013 datasets. Lastly, we conducted an experiment with LivDet 2011 and LivDet 2013 datasets to test our model. The result shows that the performance of classification of our method is superior to other methods.

Abbreviations

AC	Access Control
ACA	Average Classification Accuracy
ACE	Average Classification Error
AFIS	Automated Fingerprint Identification System
BCT	Biological Characteristics Technology
CNN	Convolutional Neural Network
DL	Deep Learning
FA	Fingerprint Authentication
FAR	False Accept Rate
FIS	Fingerprint Identification System
FLD	Fingerprint Liveness Detection
FRR	False Reject Rate
LPQ	Local Phase Quantization
ML	Machine Learning
PCA	Principal Component Analysis
PR	Pattern Recognition
SL	Supervised Learning
SPP	Spatial Pyramid Pooling
SVM	Support Vector Machine
TF	Texture Feature
USL	Unsupervised Learning

1. Introduction

With the rapid development of multimedia technology and digital image-processing technology, it becomes easy to obtain a large number of high-resolution images using sophisticated digital cameras or other high-resolution sensors devices. The availability of high-resolution images further promotes the development of image-related fields, including computer vision, pattern recognition, image processing, remote sensing technology, intelligent video analysis, intelligent transportation, and medical image diagnosis, etc. In the meantime, protecting private information of users become a core issue in various security fields.

In order to prevent attackers from breaking in the system, Access Control (AC) technology, has developed rapidly. Access control methods can mainly be divided into three categories:

➤ Prior knowledge based AC method, such as user name and static password, SMS password, PIN, etc.

➤ Trusted objects based on access control method, such as token, USB key, etc.

➤ Biological characteristics based access control method, such as fingerprint, face, signature, iris, palms, vein, etc.

Prior knowledge based AC methods and trusted objects based access control methods are easy to manage. However, they are vulnerable to eavesdropping, dictionary attack, and uncertainty factor generation. In addition, another problem about these two methods is that if the passwords and objects are not used for a long time, they are easy to be forgotten, lost and maliciously misused [1]. With distinctive advantages of convenience and security, biometrics-based AC method has been rapidly developed. Biometric recognition technology identifies an individual, based on physiological or behavioral characteristic, and matches the extracted features with those stored in advance. By using the method of feature similarity calculation, a biological measure method, called *Biometrics*, has been used.

Among various biometrics, fingerprint is unique, universal and constant, so it was probably the first biometric trait to be used to identify people. Since 1960s, Automated Fingerprint Identification Systems (AFIS) have been widely deployed in many government, law agencies and civilian applications such as ID cards, e-passports, and border control [2]. Fig. 1 presents the framework of the traditional fingerprint identification system.

The process of fingerprint identification consists of three stages, fingerprint collection stage, fingerprint verification stage, and fingerprint output stage:

➤ In the fingerprint collection stage, the fingerprint image of the users are collected and the characteristics of the fingerprints are collected and stored corresponding with the users' ID. In order to reduce computing and storage costs, before storing the users' ID and their features to the server, the fingerprint features are divided into three types according

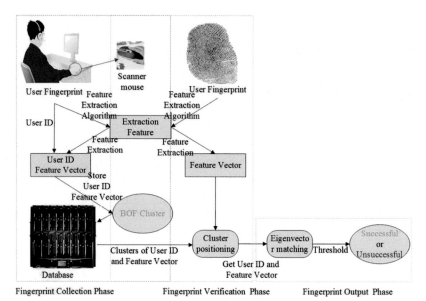

Fig. 1 The flowchart of traditional fingerprint recognition systems. *Source: Author.*

to the texture pattern of the fingerprint: bucket type, bow type and dust-pan type. Afterwards, a BOF (Bag of Feature) clustering operation is performed. The main purpose of this operation is to classify the finger-print images with similar shapes into the same cluster, which can save computing time and speed up the matching of feature vectors in finger-print verification stage.

➢ In the fingerprint verification stage, the extracted features of fingerprints are quickly located in the cluster first. Then, feature matching in the cluster is carried out.

In the fingerprint output stage, the system will give the final access right according to the threshold value. That is, if the calculated characteristic similarity is greater than the given threshold value, it indicates that the user has the right to access the system.

However, many researchers argued that the existing fingerprint authentication systems are unsafe: they are easy to be attacked by artificial replicas produced from materials such as silicon, wood glue and latex, with the help of cooperative or non-cooperative schemes [2,3]. Hence, a reliable fingerprint authentication system should be able to distinguish a spoof fingerprint from authentic ones prior to authentication. Facing this challenges, many researchers are now seeking a method that can prevent spoof attacks by artificial replicas.

The remainder of this chapter is organized as follows: in Section 2, the definition and classification of FLD are presented. In Section 3, we demonstrate how we construct our FLD model. In Section 4, we report the experimental results. Section 5 summarizes the innovative contributions of our research. In Section 6, we present the research directions in the field. Section 7 concludes our research.

2. Definition of FLD and classification of FLD methods
2.1 Definition of FLD

Although fingerprint identifier verification has become a popular and efficient security technology, most of fingerprint recognition systems are vulnerable to various threats, such as attacks at the sensor level or at the database level. Former research [4] pointed out that fingerprint sensors are easily spoofed by some artificial replicas using wax, moldable plastic, Play-Doh, clay or gelatin [5]. Putte [6] in 2000, divided the artificial fingerprint replicas into two categories: cooperation of fingerprint owner and non-cooperative of fingerprint owner. In cooperation, artificial fingerprints are created directly from alive fingerprints with person's consent using casts and molds; while in non-cooperative method, artificial fingerprints are created indirectly from latent or cadaver fingerprints on objects or bodies. Putte [6] reported that six of the fingerprint sensor spoofing incidents were caused by a collaborative approach, in which the artificial fake fingerprint was dummy silicon rubber. Fig. 2 shows the production process of two different collaborative ways to imitate fake fingerprints, and the left side of the flowchart is a collaborative approach.

Due to the above threats, there is an urgent need for a fingerprint authentication system possessing capacity of identifying artificial replicas. Many researchers have devoted to develop algorithms for systems that can resist spoofing attacks. Among these algorithms, fingerprint liveness detection (FLD) is one of the most widely used one.

The improved fingerprint identification system shall not only conduct the conventional user identity authentication, but also has the ability to verify whether the fingerprints come from alive person. To demonstrate the difference between the fingerprint recognition and fingerprint liveness detection, the workflow of fingerprint liveness detection is illustrated in Fig. 3. A comparison between Figs. 1 and 3, it reveals that the difference exists in that there is one more fingerprint liveness detection core module in Fig. 3. This core module is the research focus of this chapter.

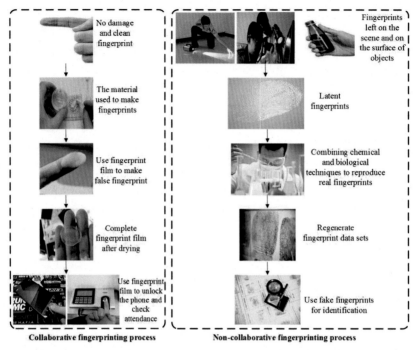

No damage and clean fingerprint	Fingerprints left on the scene and on the surface of objects
The material used to make fingerprints	Latent fingerprints
Use fingerprint film to make false fingerprint	Combining chemical and biological techniques to reproduce real fingerprints
Complete fingerprint film after drying	Regenerate fingerprint data sets
Use fingerprint film to unlock the phone and check attendance	Use fake fingerprints for identification

Collaborative fingerprinting process   **Non-collaborative fingerprinting process**

Fig. 2 Collaborative and non-collaborative processes for making fake fingerprints. *Source: Author.*

In this core module, the judgment process consists three steps. First, the image is input to the core algorithm; then, the liveness of the fingerprint authenticated is detected by using a black box test; finally, the judgment is performed to generate results. Only if the fingerprint is from an alive person, it will be sent to the next stage, i.e., the user's identification.

2.2 Classification of FLD methods

According to whether additional measurement instrument is involved, we can divide the FLD algorithms into two categories: software-based detection algorithm and hardware-based fingerprint liveness detection algorithm [7] (see Fig. 4).

Note that Fig. 4 shows the mainstream classification of FLD. Some recent studies demonstrated that anti-spoofing countermeasures based on physiologic features also can be utilized, include skin resistance, blood oxygen, pulse oximetry and other physiological liveness/vitality indicators [8]. Although these biometric properties can help recognize real and fake

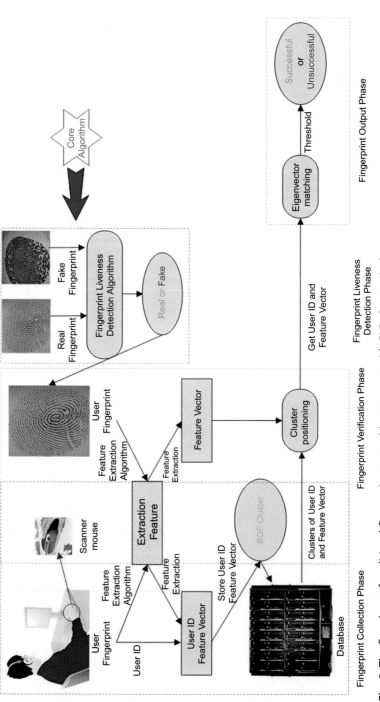

Fig. 3 The flowchart of traditional fingerprint recognition systems with FLD. *Source: Author.*

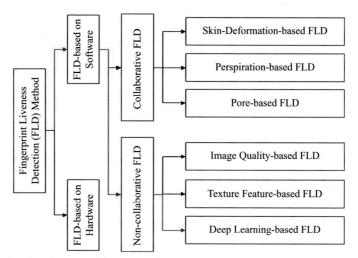

Fig. 4 The classification of fingerprint liveness detection methods. *Source: Author.*

fingerprints, the downside is that these devices are expensive and they require trained professionals to operate. In order to save costs and simplify operation, an ideal anti–spoofing detection algorithm is to use as few additional hardware or major reconfigurations as possible in existing detection systems.

2.2.1 Collaborative FLD

The cooperative fingerprint liveness detection refers to the recognition of real and spoof fingerprints with help of an alive body. That is to say, it is realized with cooperation of human being consciously. Cooperative FLD techniques are divided into skin–deformation based FLD methods, perspiration–based FLD methods and pore based FLD methods.

• Skin–deformation based FLD

Former studies show that real fingers have a significant amount of distortion than those spoof ones, when the testers press their fingers on the surfaces of fingerprint scanners. As for the real fingerprints, their degree of distortion is related to elasticity of human skin, the position and shape of finger bone, so it is difficult to imitate a real fingerprint's skin deformation. In view of this, many FLD methods based on skin deformation had been proposed in the past decades. For example, in 2006, Antonelli et al. [9] proposed a FLD method based on skin elasticity. In 2007, Zhang et al. [10] also applied skin elasticity of finger in FLD. Jia et al. [11] and Tan and Schuckers [12] used skin elasticity to discriminate the liveness of given fingerprints, and the experimental results of their approaches are satisfactory.

- Perspiration-based FLD

The Sweat glands produce perspiration, but there is no such biological phenomenon with artificial fingerprint. Therefore, the liveness of the fingerprint can be examined through analysis of sweat image. As shown in Fig. 5, fingerprints consist some ridges and valleys. Fig. 5A is an original fingerprint image, where beads of sweat are evenly along the ridges of fingerprints under microscope view; Fig. 5B, while the perspiration phenomenon is visible only in the real fingerprint image. In order to observe clearly the perspiration phenomenon, Fig. 5C shows an enlarged finger: there are many pores on the ridges of fingerprint, which are small openings on the skin from sweat glands, hair follicles, and sebaceous glands deep within the dermis. In 2003, Derakhshai et al. [8] proposed a FLD which could determine the liveness/vitality of fingerprints by observing changes of perspiration pattern on the ridges of fingers. Abhyankar and Schuckers [13–15] proposed a wavelet based FLD method using the biological phenomenon of perspiration: the authors analyzed perspiration pattern of two fingerprints collected at 0s and after 5s [14] or after 2s [13] to verify the vitality of fingerprints. In 2009, DeCann et al. [16] proposed a region labeling method to quantify perspiration of two fingerprint images captured at 0 and 1s. Region labeling method can extract trends of real and fake fingerprint images, and the fingerprints were captured using an optical fingerprint scanner, Identix DFR2100. In 2006, Tan and Schuckers [15] pointed out that the disadvantage of perspiration-based FLD was that it was time-consuming, since the users need to present their fingerprints on the scanner twice and they are not suitable for real-time authentication.

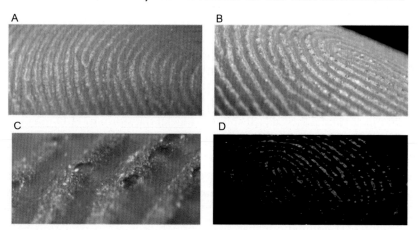

Fig. 5 Sweat fingerprint samples (Real fingerprints). (A) Original fingerprint. (B) A sweat fingerprint. (C) A magnified Sweat fingerprint. (D) Sweat fingerprint captured using sensor device. *Source: Author.*

• Pore based FLD

The eccrine sweat glands of the subcutaneous layer of the skin generate sweat ducts, where sweat ducts grow through the subcutaneous layer and dermis to the epidermis and the open conduit on the skin surface appears as a pore. The pores does not exist with fake fingerprints. In 2010, Manivanan et al. [17] developed a FLD method which could automatically extract and locate of pores in fingerprint image. In this method, only one fingerprint image rather than a series of fingerprints was used to discriminate the real fingerprints from fake ones. Nikam and Agarwal [18] extended Manivanan's method [17], and explored various thresholds value used in binarization. They detected the positions of pores for unique identification in their method, and a new image-processing algorithm using High-Pass and Correlation filtering (HCFA) technique was developed. In 2018, Yuan and Sun [19] and Kim [20] claimed that the number of intrinsic features of friction ridge skin (pores) could be used as a liveness sign to detect the presentation attacks. Jia et al. [11] developed a FLD method analyzing pores location distribution of two images captured at time 0 and 5 s. Another similar study can be found in [1].

2.2.2 Non-collaborative FLD

Non-cooperative FLD is to examine the fingerprint image without conscious cooperation with human being. It has a wider application scope than the cooperative FLD. Non-cooperative FLD can be divided into three categories: image quality based FLD, texture feature based FLD and deep learning based FLD.

• Image quality based FLD

The image quality of artificial fingerprints is always worse than that of the real on analysis of higher resolution fingerprint images. This method relied on the fact that the surface of fake fingerprints is coarser than real fingerprints, because the materials for producing fake fingerprints are composed of large organic molecules that often come together during processing. In 2008, Nikam and Agarwal [18] applied ridgelet transform to FLD to extract texture information using only one fingerprint image. The authors observed that the texture information of a fake fingerprint was different from a real fingerprint. In addition, feature dimensionality reduction was performed to solve the over-fitting problem caused by insufficient image sets using Principal Component Analysis (PCA) algorithm. In 2006, Yuan and Galbally [2,21] pointed out that, even though previous perspiration-based FLD methods were efficient and effective against spoofing attacks, they still have trouble that more than one image were required to implement the detection. Therefore, they were not suited for on-line processing.

• Texture feature based FLD

Texture is a significant visual clue that reflects the homogeneity phenomenon of an image. Different from image features such as grayscale and color, texture is represented by grayscale distribution of pixels and surrounding spatial fields. In addition, texture features are not based on the features of pixels, but need to be calculated statistically in an area containing more pixels. In pattern recognition, even when the pixel block is partially offset, the features can still be matched. Abhyankar and Schuckers [22] proposed a FLD method combining the multiresolution texture analysis and the interridge frequency analysis. Martins et al. [23] assessed the recently introduced Local Phase Quantization (LPQ) algorithm and applied it to fingerprint liveness detection. A series of experiments based on several common feature extraction algorithms were performed, including LPQ [2], LBP, LBP's variants [24], and results also demonstrated that LPQ based detection rate was the best.

• Deep learning based FLD

Most aforementioned FLD methods are based on feature engineering, in which handcrafted features representations mainly rely on the experience and professional knowledge. Therefore, to improve precision of FLD, how to extract more effective feature descriptors is a key issue (see Fig. 6 [2]).

In response to this issue, application of Deep Learning technology in FLD is developing fast. In the recent 6 years, Deep Learning, combining low-level features to form more abstract high-level representation attribute categories or features to discover distributed feature representations of data, quickly becomes popular in various research fields.

Convolutional Neural Network (CNN) is one of the most widely used supervised learning models in Deep Learning. Because CNN is capable of characterizing self-learning, it is widely used in objects classification, detection and segmentation. In the structure of CNN, convolutional layer and pooling layer are the most basic components and the core part. The characteristics of the high level are the combination of the characteristics of the low level, and the characteristics from the low level to the high level are more and more abstract and can express semantics or intentions. Convolutional layer is similar to a feature extractor.

Nogueira first proposed a FLD method based on CNN in [25]. This method was not an end-to-end liveness detection method. A convolutional neural network was viewed as a feature extractor. In this method, two kinds

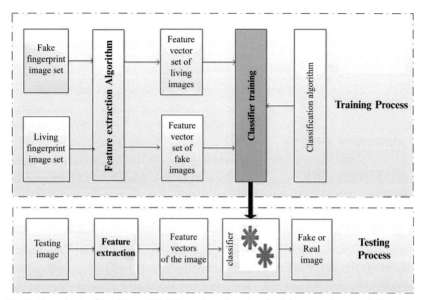

Fig. 6 Flowchart of traditional FLD based on handcrafted feature. *Source: Author.*

of features were extracted separately using convolutional networks with random weights and a classic local binary pattern. After feature extraction, four pipelines were utilized in their experiments, and data augmentation had been performed to increase classifier's performance. No matter which feature was followed by a Support Vector Machine (SVM) classifier. Four pipelines are including CN + PCA + SVM, LBP + PVA + SVM, AUG + LBP + PCA + SVM, and AUG + CN + PCA + SVM. The experiments were performed using three public fingerprint datasets LivDet 2009, 2011 and 2013. In 2016, Nogueira et al. [26] applied convolutional neural networks (CNNs) to fingerprint liveness detection. Experiments were evaluated on three public data LivDet 2009, 2011, and 2013, which composed of 50,000 real and fake fingerprints images. In this method, four different models were used to compare the detection performance: CNN pre-trained on ImageNet, CNN with random weights, CNN fine-tuned with the given fingerprint images, and a local binary pattern method. To enhance the classification performance, data augmentation was performed. Using these pre-trained CNN, they obtain good performance in small training sets (400 samples). Experimental results indicated that pre-trained CNNs could obtain classification results without requiring architecture or hyper parameter selection. Meanwhile, pre-trained CNNs showed optimal generalization performance in cross–dataset experiments than these methods without using CNNs. In the fingerprint liveness detection competition 2015, this method won the first prize with an overall accuracy of 95.5%. In 2017, Yuan et al. [27] pointed out that most current FLD approaches are focused on the construction of handcrafted features. These approaches do not take into account the spatial information of image pixels. To solve this problem, Yuan proposed a feature extraction method concatenating low-level edge and shape features from a large amount of labeled data. To reduce the redundant features, Region of Interest (ROI) and Principal Component Analysis (PCA) operations were performed for features learned of each convolutional and pooling layer. Afterwards, the processed features were fed into Support Vector Machine (SVM) classifier. The experimental results showed that the classification performance of their method was superior to other previous methods.

3. Construction of our FLD

3.1 Rationale for our construction

We notice that all the aforementioned detection algorithms are using hand-crafted features representations, which mainly rely on the experience and professional knowledge. As shown in Fig. 6, the traditional FLD contains two processes: Training Process and Testing process. Whether in the training process or in the test process, the feature extraction of the image is the most critical part, that is to say, the traditional fingerprint activity detection performance depends on the feature engineering extraction. In order to improve precision of FLD, developing effective feature descriptors is a key issue. Moreover, due to loss of spatial location information and lack of considerations for the details of real fingerprint images, it is difficult to achieve a balance between discriminatingly and robustness of methods. Features extraction methods based on deep learning demonstrate tremendous potential to address these issues.

Meanwhile, big date brings more challenges on deep learning strategies for object classification and analysis. As shown in Fig. 7A–C, the scales of fingerprint samples often vary for different fingerprint sensor devices. Traditional CNNs [4] do not fully use the scale information, because they only extract the high-level features of fingerprint images from a fixed scale (e.g., 224×224). Additionally, the convolutional operation and pooling operation are no limit to the size of input image in the forward propagates of CNNs, and these operations can only affect the size of features maps of convolutions and pooling layers. However, the fully connected layers restrict the size of input images, since these operations need to specify the

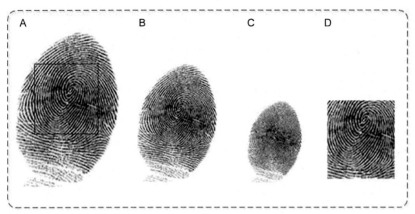

Fig. 7 Fingerprint images in different size/scale. (A) The size of image is 312×372, (B) the size of image is 156×186, (C) the size of image is 78×93, and (D) is the image after crop operation. *Source: Author.*

number of the input layers neurons and the number of the output layers neurons in full connection layer. To cope with the different size of fingerprint images, two common operations have been explored: crop and warp. Although they solve the problem of image scale, the former operation can cause loss of some important edge information [as shown in the red window in Fig. 7A. The features descriptors of latter operation are very susceptible to image deformation (as shown in Fig. 7B). Thus, classification performance is compromised due to lack of effective discriminant information.

To address the above issues, we attempt to construct multiscale CNNs with fine-tuning. CNNs with Spatial Pyramid Pooling network (SPP-net), which was proposed in [28], are used to settle the problem of different image scales in our method. By using an SPP-net layer instead of the general pooling layer between the last convolutional layer and the first fully connected layer, our method can generate fixed-length vector expressions for images of any scale without any crop or warp operation. In addition, our model is also robust to fingerprint image deformation [20]. During training model, fine-tune the model parameters of full connection layers can strengthen generalization performance of the trained model. Because fully connected layers can generate a plenty of parameters, the CNNs inevitably result in over-fitting problem. We therefore use a pre-trained network model as parameters initialization of our model. The fingerprint testing samples with labels are verified using the trained network model.

3.2 Structure of our FLD method

Our FLD scheme contains five components: Input, Convolution, SPP-net, full Connection and Output. The structure of it is illustrated in Fig. 8.

3.2.1 Convolutional layers and pooling layers

The convolutional layer operations are regarded as the high-level feature representations of the given fingerprints by finding the correction from raw pixel level intensity. Drawing on [29], we construct an improved CNNs with SPP-net for FLD. As shown in Fig. 9, (A) is our model structure, (B) is the Alexnet model architecture. In both structure, the output of each layer is entered as the input of the successive layer. For simplicity, we only show difference between the two model architectures. Different from the former method in [26], the sequence of the normalization layer and pooling layer has been adjusted to reduce the parameters of trained model in our scheme and to eliminate the image size problem.

In the CNNs, alternating convolutional layers, pooling layers, full connection layers, and a final classification layers are included. In Fig. 10, we visualize a feature map using a convolutional operation and the pooling operation. Convolutional features are represented through computing the inner product of original fingerprint image and filters, and the process of convolution is considered as the process of feature extraction. Next, ReLU is viewed as the activation function to compute feature maps. After the convolution, max-pooling operation is performed to reduce the dimensionality of feature maps and prevent over-fitting. The principle of max-pooling counts the maximum in the sliding windows. Such as the green solid line window in Fig. 10, all the convolutions are followed a non-linear activation operation ReLU. Model takes a three-channel fingerprint image as input. The first convolution operation generates 64-channel feature map with spatial dimension of 224×224. Then, the 2D maximum pooling operation reduces the spatial dimension to 112×112.

3.2.2 The construction of SPP-net deep feature vector

The scale of fingerprints images is various, due to different fingerprint acquisition devices; different scale contains different detail information of the images. Our scheme attempts to learn multiscale high-level features, so as to capture detail spatial information. To achieve this goal, a fixed-length

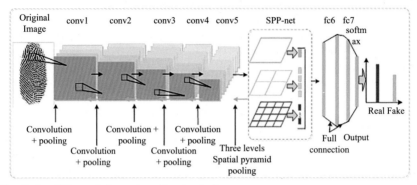

Fig. 8 Our FLD model structure. *Source: Author.*

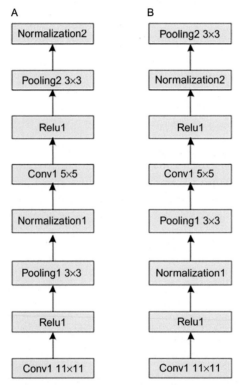

Fig. 9 Comparison of our model structure and the original model architecture: (A) the part of our model structure and (B) the part of original model architecture. *Source: Author.*

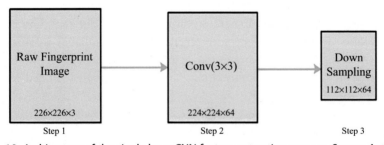

Fig. 10 Architecture of the single layer CNN feature extraction process. *Source: Author.*

vector set is necessary for the classifiers or full connection layers. Therefore, we establish a network layer structure based Spatial Pyramid Pooling network (SPP-net).

The SPP-net keep spatial information by pooling in the local spatial sub-regions. Features maps of the last convolutional layers are partitioned into sizes proportional to sub-regions, so the number of sub-regions is fixed for images of an arbitrary size/scale. As shown in Fig. 8, in our scheme, following five successive convolutional layers and two successive general pooling layers, a spatial pyramid pooling layer is installed. It contains two full connection layers, and a classification layer. Afterwards, to eliminate the image size problem, the spatial pyramid layer has been added between the last convolution layer and the first full face layer.

Suppose the size of the last convolutional layer output image (feature map) is $a \times a$, each feature map is divided into $n \times n$ blocks. SPP-net is viewed as convolution operation, where the size/scale of sliding window is $\text{win} = \lceil a/n \rceil$ and the stride is $\text{str} = \lfloor a/n \rfloor$, where $\lceil . \rceil$ and $\lfloor . \rfloor$ are ceiling operator and floor operator, respectively. Three layers pyramids are used to extract high-level semantic features, in which the divided sub-regions are set, respectively, $n \times n$ as 1×1, 2×2 and 4×4. In each spatial sub-region, SPP-net is used to pool the responses of each convolutional kernel. The final output is composed by concatenating three layers pooling results to generate a fixed-length semantic representation Km for input image of arbitrary size/scale, in which m is the number of sub-regions and K is the number of the features maps in the last convolutional layers. After this process, the fixed-length vectors are fed into the first full connection layer.

3.2.3 Fine tuning fingerprint neural network parameters

Limited by the number of fingerprint images, trained model classifiers are prone to over-fit. In order to resolve this problem, we obtain some parameters from pre-trained model using ImageNet 2012 database. Then we use the training samples from LivDet databases to fine-tune the pre-trained model parameters. The training procedure of our network model is based on the open source Caffe DCNN library [30,31]. After the pre-training using ImageNet 2012, the weights and bias parameters learned in the five convolutional and general pooling layers are viewed as the initialization parameters of our network model. The pre-training model parameters are directly used as the initialization of our model, and we only need to use the training samples of fingerprint data set to fine-tune our model parameters.

4. Experimental results

In order to examine the detection accuracy of our ICNNSPP, the data sets of LivDet 2011 [1] and LivDet 2013 [32] from 2011 and 2013 Fingerprint Liveness Detection are used. We compare our experimental results with several other existing algorithms.

In this section, we first make a brief description of the two public databases and experimental setup in this section. We then explain performance evaluation criteria and experimental requirements. Lastly, some numerical experiments on two well-known benchmark data sets are performed to demonstrate the effectiveness of our algorithm.

4.1 Experimental setup and datasets

Our operating environment is based on the open source code of cuDNN and Caffe framework. The operating system is Linux Mint 18 version, and all experiments are all implemented using python 3.5.2 programming on a single GeForce GTX 1080 GPU (8G memory) with 2 days. The detailed configuration of operating environment is listed in Table 1. The experimental setup is listed in Table 1.

The Department of Electrical and Computer Engineering of the Clarkson University and the Department of Electronic Engineering of the University of Cagliari [32] held the first FLD competition (LivDet) in 2009, and the competition is conducted every 2 years since then. The purpose of FLD competition is to provide academic and industrial institutions with benchmark data sets for comparing algorithms' results. We choose datasets from LivDet 2011 and LivDet 2013 to evaluate our scheme.

The LivDet 2011 database, contains 16,056 fingerprints of both real and fake using four different flat optical sensors. Two types of fingerprint samples

Table 1 The requirements of our experimental operating environment.

Hardware condition	Software condition
CPU: Intel@ Core i7-6850k 3.60GHz x 12	Operating System Version: Ubuntu 16.04
Memory: 64GB	Run Environment: Python 3.5.2 + Cuda 8.0 + Tensorflow 1.3.0
Graphics: NVIDIA@ GeForce GTX 1080 Ti/PCle/SSE2	Compiler Environment: Sublime Text

Source: Author.

Table 2 The distribution of the LivDet2011 and LivDet2013 datasets.

Dataset ID	Sensor	Res. (dpi)	Image size	Samples in training set		Samples in testing set	
				Live	Spoof	Live	Spoof
Liv2011–1	Biometrika	500	315×372	1000	1000	1000	1000
Liv2011–2	Digital	500	355×391	1004	1000	1000	1000
Liv2011–3	Italdata	500	640×480	1000	1000	1000	1000
Liv2011–4	Sagem	500	352×384	1008	1008	1000	1036
Liv2013–1	Biometrika	569	352×384	1000	1000	1000	1000
Liv2013–2	CrossMatch	500	800×750	1250	1000	1250	1000
Liv2013–3	Italdata	500	480×640	1000	1000	1000	1000
Liv2013–4	Swipe	96	1500×208	1221	979	1153	1000

Source: LivDet2011 and LivDet2013.

are included in LivDet 2011: training Set with a total of 8020 images and a total of 8036 images in Testing Set. Two types of fingerprint samples datasets are included in LiveDet 2013: training dataset with a total of 8450 images and Testing Set with a total of 8403 images. The LivDet 2013 [32] dataset consists of total 16,853 real and fake fingerprints. The detailed distribution of the LivDet 2011 and LivDet 2011 is presented in Table 2. Fig. 11 lists some fingerprint samples from four different optical sensors.

During the process of pre-training parameters, the dataset we use is ImageNet 2012 dataset. ImageNet 2012 data set is composed of total 1,419,722 images and divided into 1000 classes, which are collected according to the WordNet hierarchy. Some samples from ImageNet 2012 are shown in Fig. 12.

4.2 Performance evaluation

We use Average Classification Accuracy (ACE) [26,33] to evaluate and compare detection performance. The evaluation metrics of ACE is formulated as:

$$ACE = \frac{FAR + FRR}{2} \tag{1}$$

where in Eq. (1), False Accept Rate (FAR) is the percentage of misclassified real fingerprints and FRR (False Reject Rate) is the percentage of misclassified as fake ones. FAR and FRR can be expressed as:

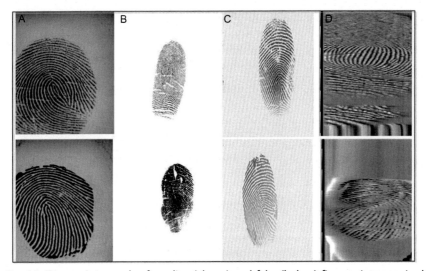

Fig. 11 Fingerprint samples from live (above) and fake (below) fingerprints acquired with four different sensors: (A) Biometrika, (B) Crossmatch, (C) Italdata, and (D) Swipe. *Source: Author.*

$$FAR = \frac{misclassified\ real\ fingerprint}{total\ number\ of\ real\ fingerprint} *100 \qquad (2)$$

$$FRR = \frac{misclassified\ fake\ fingerprint}{total\ number\ of\ fake\ fingerprint} *100 \qquad (3)$$

The value of ACE ranges from 0 to 100. In addition, the smaller the ACE is, the better the performance is. Similarly, the Average Classification Accuracy (ACA) is used to measure the correct classification performance of the algorithm, and the calculation formula can be obtained according to the average classification error rate, whose solution formula is:

$$ACA = 1 - ACE \qquad (4)$$

4.3 Experimental process and results

The pre-trained parameters, which taken as the parameter initialization of our weights and bias, are learned by using ImageNet 2012 dataset. Then we fine-tune our model parameters of the last two fully connected layers using LivDet 2011 and LivDet 2013 datasets. Due to the diversity of image expansion methods, it is difficult for us to verify whether the detection

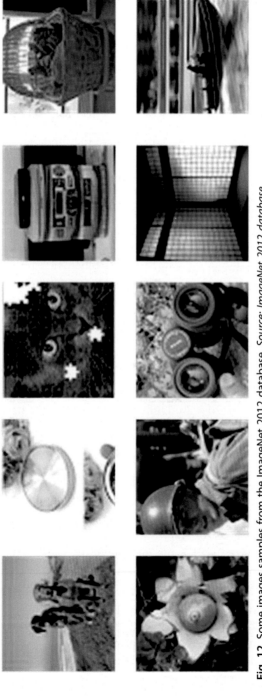

Fig. 12 Some images samples from the ImageNet 2012 database. *Source: ImageNet 2012 database.*

performance of fingerprints is caused by the image expansion or the proposed model structure. Different from the work [26], no image expansion algorithm was performed in our algorithm.

For simplify, convolutional layers, spatial pyramid pooling layers and full connection layers are abbreviated as conv, SPP and fc, respectively. For example, conv4 represents the fourth convolutional layer; fc7 represents the second fully connected layer. After performing the five convolutional and two max–pooling operations, the conv5 contains rich high–level semantic information of fingerprints. To eliminate the limitation of images sizes, the last pooling layer in our method has been replaced in a SPP–net layer, which can make any image size as the input of the model, and the red dotted line window shows the key part of our network model in Fig. 8. In our method, three layers pyramids are used to learn high–level structural and semantic feature representations. Due to the diversity of images sizes (as shown in Table 2), supposed that the size of input fingerprint image is 355×355, the parameters of conv1 layer are filter size is 11, stride is 4, therefore, the size of output image is $((355 - 11)/4 + 1) \times ((355 - 11)/4 + 1) = 86 \times 86$. Pooling operation adopts the method of taking the maximum value in the field of 2×2, and the size of filter is 3×3, stride is 2. After pooling operation, the size of output image is 43×43. Note that if $(86 - 3)/2$ indivisible, then the outermost edge of the image is filled with 0. The parameters of conv2 are that pooling layer filter size is 3×3 and stride is 2, and then the size of output image is 20×20. The size of conv3 is 20×20, and the sizes of conv4 and conv5 are 20×20. Different from [26], SPP–net layer replaced the last pooling layer. Since the size of output of the last conv5 layer is 20×20, the size of sliding windows are 20, 10 and 5, respectively, and strides are 20, 10, and 5, respectively, when $n \times n$ represents 1×1, 2×2, and 4×4, respectively. The calculation about the size of sliding window and stride has been analyzed in X.3.2.1. Afterwards, the final output feature vectors are composed of concatenating three layers pooling results generating a fixed-length feature vector $Km = 256 \times 21$, in which $m = 21$ is the number of sub-regions and $K = 256$ is the number of the features maps or filters in the last convolutional layers. The fixed-length vectors are considered as the input of the first full connection layer. Finally, the extracted features are fed into softmax classifier. After two processes of forward propagation and back propagation, the final model parameters are learned using supervised learning method, and the performance of the model is verified by testing fingerprint datasets and we can obtain the probability of a certain type of fingerprint.

Table 3 The summary of the models in experiment.

Experiment ID	ModelName	Pipeline	Description
1	MSP-Improved CNN	Convolutional Layers + Max pooling + normalization + SPP-net + Full Connection	Features extracted based on five Convolutional Layers, then normalization operation, SPP-net is introduced to address the multiscale issue
2	MSP-Improved CNN with Fine-tune	Fine-tuned model + Convolutional Layers + Max pooling + normalization + SPP-net + Full Connection	Features extraction, pooling operation, SPP, and classifier learning are all based on Pre-trained model using ImageNet 2012
3	CNN-Improved Alexnet	Convolutional Layers + Max pooling + normalization + Full Connection	Features extracted based on five Convolutional Layers, and adjust the order of pooling and normalization operations different from [26]

Source: Author.

Three sets of experiments (MSP-Improved CNN, MSP-Improved CNN with Fine-tune, and CNN-Improved Alexnet, as marked in Table 3) were carried out. Different from the previous work [26], there is no limitation on the size/scale of the fingerprint images in Experiment ID 1 and Experiment ID 2. Thus, our method keeps the spatial information and prevents image deformation. In Experiment ID 2, pre-trained model parameters using the auxiliary ImageNet 2012 database are performed. Then we fine-tune our fingerprint network structure based on the training samples. Despite they are different from fingerprint images, pre-trained operation using the ImageNet image does help improve the detection performance. The pre-trained procedure of our experiments is conducted by the open source Caffe DCNN library. In order to represent high-level structural features from raw pixels and train the classifier model quickest possible, GPU device is used in our framework. After finishing the pre-training operation, the weights and bias parameters learned are transferred to the

Table 4 Average classification errors (ACE) of our three experiments.

Experiment ID	ACE (%) in LivDet 2011 database	ACE (%) in LivDet 2013 database
1	5.548	2.048
2	4.868	1.94
3	6.18	2.22

Source: Author.

Table 5 Classification error ratio of different algorithms with LivDet 2011 dataset.

	The average classification error (ACE) (%)				
Methods	Biometrika	Digital	Italdata	Sagem	Average
Our method	**7.6**	**2.1**	11	**2.5**	**5.65**
Winner [36]	20	36.1	21.8	13.8	22.9
Pore Analysis [37]	26.6	31.4	23.4	22.0	25.9
MBLTP [38]	9.7	7.0	16.0	5.8	9.62
Fusion [37]	18.4	15.2	7.8	6.7	12
LASP [39]	22.6	27.1	17.6	17.58	21.22
Baseline [37]	20.6	14.0	8.4	8.4	12.9
SURF + PHONG [40]	8.76	6.9	**7.4**	6.23	7.32
WLD [17]	13.3	13.8	27.7	6.7	15.4

The value in the table is in comparison with each other. The smaller the value is (boldface), the better this algorithm functions as explained in the paper.
Source: Author.

fingerprint images and kept fixed. The ACE values of our three models on are listed in Table 4. It shows that the ACE of Experiment ID 2 is optimal within our three experiments.

To show the advantages of our method, several different experiments on the two public fingerprint datasets are performed and compared with the existing methods, including CNN [26], LCP [20], MSDCM [34], DBN [35], and LLF-SA [35]. Tables 5 and 6 list the detailed comparison results. We can find that no matter for LivDet 2011 fingerprint dataset or LivDet 2013 fingerprint dataset, ACE performance of our proposed algorithm is the best. The optimal results for each fingerprint sensor are shown in bold in each column. The result of Italdata and Swipe on LivDet 2013 dataset is

Table 6 Classification error ratio of different algorithms with LivDet 2013 dataset.

Methods	The average classification error (ACE) (%)				
	Biometrika	CrossMatch	Italdata	Swipe	Average
Our method	2	7.42	**0.85**	**0.74**	**2.75**
PHONG [40]	3.87	9.92	6.7	9.05	7.24
CN [41]	4.55	**5.2**	47.65	5.97	15.84
ULBP [32]	10.68	46.09	13.7	14.35	21.21
SURF [40]	5.75	6.08	4.6	4.6	5.26
MSDCM [42]	3.55	20.84	2.35	5.25	7.59
HIG DBP [43]	3.9	28.76	1.7	14.4	12.19
Winner [1]	4.7	31.2	3.5	14.07	13.37

The value in the table is in comparison with each other. The smaller the value is (boldface), the better this algorithm functions as explained in the paper.
Source: Author.

Table 7 Overall performance with the LivDet 2011, 2013 datasets.

Methods	MSDCM [34]	CNN-VGG [26]	CNN-Alexnet [26]	Ours without SPP-net	Ours with SPP-net
LivDet2011	20	36.1	21.8	13.8	22.9
LivDet2013	26.6	31.4	23.4	22.0	25.9
Average	13.3	13.8	27.7	6.7	15.4

Source: Author.

close to 0.7 in Table 6, and the ACE of our algorithm is still 1.67% and 2.51% lower than two results of the second place [24]. In our approach, the results of Digital, Sagem, Italdata, Swipe devices are better. In general, our method is more suitable for application of FLD.

The overall performances of the different combination methods (with SPP-net or without SPP-net) are reported in Table 7. Compared with [8], the average performance of ours with SPP-net is obviously superior to two methods in [26]. And the results with LivDet 2013 of our improved network model (ours without SPP-net) is also better than that of 2013 of [26].

Table 8 demonstrates the average training time and testing time on the LivDet 2011 and LivDet 2013. By observing Table 8, we found that the testing time of a fingerprint image on two different public fingerprint

Table 8 Average training and testing time.

Database	Training all fingerprints with GPU	Ours with SPP-net testing a fingerprint (single CPU with GPU)
LivDet2011	11–12 h	226 ms
LivDet2013	8–9.5 h	208 ms

Source: Author.

data sets is close to 10 h, which is not suitable for practical application due to the large number of training data sets. After learning these model parameters, classification performance based on testing samples has been performed, and the testing time of a fingerprint image is close to 210 ms, which is satisfactory and acceptable in practical application.

In order to compare the performance of different fingerprint sensors with pre-trained model and those without pre-trained model, we computed the Average Classification Accuracy (ACA) with regard to different finger-print sensors in LivDet 2011 and LivDet 2013 datasets. We also plot the corresponding values of different fingerprint sensors with pre-trained model or without pre-trained model in Fig. 13. Fig. 13 shows that the ACA of pre-trained models, which can be obtained according to the formula (4). That is to say, each iteration is 1000 times, the value of ACA is computed by formula (4). By observing Fig. 13, the pre-trained classification accuracies are obviously superior to results of those non-trained ones. Moreover, as the number of iterations increases, the overall ACA will continue to increase until the trend level no longer increases. Taking CrossMatch sensor as an example in LivDet 2013 fingerprint images dataset, ACA value before pre-training was close to 90%, while after pre-training, it was close to 96%.

5. Summary of contributions of our research

The major contributions of our research can be summarized as follows:

1. In our FLD scheme, a SPP–net is employed to remove the restrictions on fixed size or fixed scale of images by CNNs. This progress can reduce the information loss caused by crop and warp operation.
2. Our model can automatically learn the high-level structural features using an improved CNN model from large amounts of labeled data.

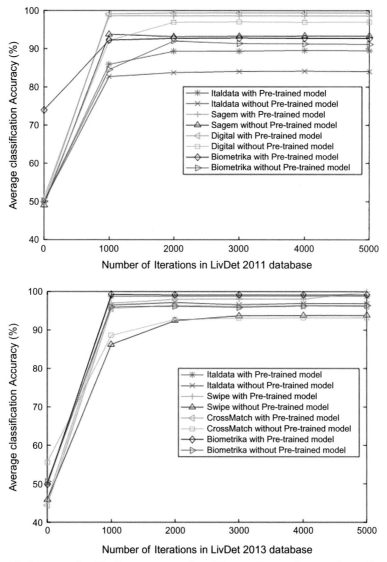

Fig. 13 Average classification accuracy for different fingerprint sensors with and without pre-trained model. *Source: Author.*

In addition, features representations have scale invariance from three scales pyramids in a hierarchical manner, so our method is more robust to image deformation.

3. In order to solve the over-fitting issues, the pre-trained model parameters based on ImageNet are adopted and used as initialization of our

model parameters. It greatly accelerates the training process and improves the classification performance. Moreover, using three layers pyramids structure in the SPP-net simplifies the complexity of the model, and reduces the learning parameters of model.

6. Research directions in the field

There are many research directions in the FLD filed [44–46]. FLD can be categorized as two patterns of recognition, that is, test of the fingerprint from a living body, or from a non-living body. The key to realize a high-performance FLD algorithm is how to extract and construct a feature representation that distinguishes between true and false fingerprints. At present, the existing FLD algorithms mainly fall into two types, one is based on features obtained by manual feature engineering, and the other is based on the study of high-level semantic features by deep learning. In the area of research based on features obtained by manual feature engineering, researchers need to possess a large quantity of techniques and programming skills. How to solve the memory shortage, data overflow, model overfitting, parameter selection and optimization are problems researchers need to focus on; at the same time, how to better interpret the extracted features is another challenge that researchers need to focus on. In the of research based on the study of high-level semantic features by deep learning, deep learning technology can automatically learn high-level semantic features from big data, which can better tackle the problems that are difficult to resolve in the research area based on manual feature engineering [46–49]. In order to distinguish between true and false fingerprints, how to design a deep learning model is the focus. In addition, because the deep learning technology has the characteristics of self-learning, whether the deep learning model can be used as a feature extractor and can be combined with the existing classifier for FLD is also a research direction [50–53].

On our research agenda, we will continue to study the corresponding detection performance under deeper layers by increasing the number of layers of in our model. In addition, when the amount of data is large, it is not practical to manually label the image tags, therefore, how to design an unsupervised FLD [54–56] algorithm using unsupervised learning is another task within our future research.

7. Conclusion

No matter for image classification or object detection or segmentation, feature extraction is the key to achieve a higher classification or detection performance. Most of the existing most features representations algorithms are based on handcrafted features, which mainly rely on the experience and professional knowledge. Recently, the original deep learning, especially CNN model, has attracted a lot of attention in the field of pattern recognition and computer vision. Although CNN model can automatically learn high-level semantic information from labeled data, CNNs require a fixed-size image as the input to model in previous approaches. In order to solve the problem of fixed-size, two common operations are introduced, including crop operation or warp operation, which will affect final classification performance due to loss of some discriminative spatial information. In this chapter, to eliminate the limitation of input images size, a novel CNN model framework with SPP-net is proposed to detect the fingerprint liveness. In addition, high-level structural features based on multiscale images are learned, so our method has the scale invariance and more robust to scale invariant. Weights and bias parameters learned in pre-trained network model using ImageNet 2012 database are migrated to our model, which are viewed as the parameter initialization of our model architecture. Then, we fine-tune our SPP-net by employing the training samples from LivDet 2011 and LivDet 2013 databases. Our experiment proves that the classification performance of our model is better than other methods and is more suitable for FLD; it can better prevent spoofing attacks by artificial replicas.

Acknowledgments

The authors are grateful for the anonymous reviewers who made constructive comments and improvements for our manuscript. This work is supported by the National Key R&D Program of China under grant 2018YFB1003205; by the National Natural Science Foundation of China under grant U1836208, U1536206, U1836110, 61602253, 61672294; by the Jiangsu Basic Research Programs-Natural Science Foundation under grant number BK20181407; by the Canada Research Chair Program and the NSERC Discovery Grant; by the Priority Academic Program Development of Jiangsu Higher Education Institutions (PAPD) fund; by the Research Startup Fund of NUIST (1441102001002); by the Collaborative Innovation Center of Atmospheric Environment and Equipment Technology (CICAEET) fund, China.

Key terminology and definitions

Access Control (AC) AC is a technique that restricts user access to certain information items by user identity and a defined group to which they belong, or restricts the use of certain control functions. Access control is typically used by system administrators to control user access to network resources such as servers, directories, files, and so on.

Convolutional Neural Network (CNN) CNN is one of the representative algorithms of deep learning. Since the convolutional neural network is capable of shift-invariant classification, it is also called "Shift-Invariant Artificial Neural Networks (SIANN)." The study of CNN began in the 1980s and continued in 1990s. The time delay network and LeNet-5 were the earliest CNNs. After the 21st century, with the introduction of deep learning theory and numerical calculations, CNN has become hotspot in research and application in the fields of computer vision and natural language processing.

Biological Characteristics Technology (BCT) Biometric Identification Technology refers to technology that use human biometrics for identity authentication. It is combined with high-tech means such as optics, acoustics, biosensors, and biostatistics, and uses the inherent physiological and behavioral characteristics of the human body to identify individuals.

Fingerprint Authentication (FA) The uneven texture on the front side of the finger contains a lot of information. The lines of these skins are different in patterns, breakpoints, and intersections. They are "features" in information processing, and these features are unique and permanent, and can be used to verify people's identity.

Fingerprint Identification System (FIS) A FIS is a system that uses biometric technology to store digital imagery of individual fingerprints for database comparison to produce a match. Fingerprints are considered a foolproof method for identification purposes because each fingerprint is unique. As digital technology progresses, fingerprinting is increasingly being used as a fraud prevention measure. This type of technological safeguard may be stored with personal data, such as passwords and personal identification numbers (PIN).

Fingerprint Liveness Detection (FLD) Biometric authentication systems in general, and FIS in particular, can be deceived by presenting fake samples of the biometric trait to the sensor used in a specific system. The easiest way of creating a fake sample is by printing the fingerprint image into a transparent paper, and a more successful method is to create a 3D fake model with the fingerprint stamped on it. FLD techniques and alteration detection methods are both methods included in the presentation attack detection (PAD) methods. FLD is based on the principle that various of minutiae points, highlighting the difference between live and fake fingerprints, can be recorded using a standard scanner devices, namely, FLD is viewed as a binary classification problem. Until now, extensive research efforts have been dedicated to achieving the security of fingerprint liveness detection for the identity authentication application, and these approaches are mainly grouped into two categories through additional hardware or by processing the obtained image: hardware-based and software-based approaches.

Deep Learning (DL) DL is a new field in machine learning research. Its motivation is to build and to simulate a neural network for human brain analysis and learning. It mimics the mechanism of the human brain to interpret data such as images, sounds and texts. DL methods can be categorized as supervised learning and unsupervised learning. For example, Convolutional Neural Networks (CNNs) is a DL model under supervised learning; Deep Belief Nets (DBNs) is a DL model under unsupervised learning.

Machine Learning (ML) ML is a general term for a class of algorithms that attempt to mine the implicit rules from a large amount of historical data and use them for prediction or classification. More specifically, ML can be viewed as looking for a function, and input is sample data. The output is the desired result, but this function is too complicated to be formally expressed. It is important to note that the goal of ML is to make the learned functions work well for "new samples," not just for training samples. The ability of the learned function to apply to a new sample is called generalization capability.

Supervised Learning (SL) SL is a machine learning task that infers a function from the labeled training data. The training data includes a set of training examples. In SL, each instance consists of an input object (usually a vector) and a desired output value (also known as a supervisory signal). The SL algorithm analyzes the training data and produces an inferred function that can be used to map out new instances. An optimal solution would allow the algorithm to correctly determine the class labels of those invisible instances. This requires learning algorithms to be formed in a "reasonable" way from one training data to invisible.

Unsupervised Learning (USL) USL is a branch of machine learning that learns from test data which has not been labeled, classified or categorized. The data given to unsupervised algorithm are not labeled, which means only the input variables(X) are given with no corresponding output variables. In unsupervised learning, the algorithms are left to themselves to discover structures in the data.

Texture Feature (TF) A texture can be thought of as a set of primitive texels in a particular spatial relationship. Thus, texture is an important low-level feature that can be used to describe the contents of an image or a region and help to segment images into regions of interest and to classify those regions. Image textures can be artificially created or found in natural scenes captured in an image. For more accurate segmentation, the most useful features are spatial frequency and an average gray level. There are two approach to analyze an image texture in computer graphics: the Structured Approach and the Statistical Approach.

Pattern Recognition (PR) PR is an automated recognition of patterns and regularities in data. PR is closely related to AI and ML applications, such as data mining and Knowledge Discovery in Databases (KDD). ML is one approach to PR, while other approaches include handcrafted (not learned) rules or heuristics.

References

[1] D. Yambay, L. Ghiani, P. Denti, G.L. Marcislis, F. Roli, S. Schuckers, LivDet 2011-fingerprint liveness detection competition 2011, in: Iapr International Conference on Biometrics, IEEE, 2012, pp. 208–215.

[2] C. Yuan, X. Sun, R. Lv, Fingerprint liveness detection based on multiscale LPQ and PCA, China Commun. 13 (7) (2016) 60–65.

[3] X. Li, W. Cheng, C. Yuan, W. Gu, B. Yang, Q. Cui, Fingerprint liveness detection based on fine-grained feature fusion for intelligent devices, Mathematics 8 (4) (2020) 517.

[4] B. Tan, S. Schuckers, Liveness detection using an intensity based approach in fingerprint scanners, in: Processing of Biometric Consortium Research Symposium, 2005.

[5] S.A.C. Schuckers, Spoofing and anti-spoofing measures, Inf. Secur. Tech. Rep. 7 (4) (2002) 56–62.

[6] T. Putte, J. Keuning, Fingerprint recognition don't get your fingers burned, in: Fourth Working Conference on Smart Card Research and Advanced Applications, Kluwer Academic Publishers, 2002, pp. 289–303.

[7] Z. Xia, C. Yuan, R. Lv, X. Sun, N.N. Xiong, Y. Shi, A novel weber local binary descriptor for fingerprint liveness detection, in: IEEE Transactions on Systems, Man, and Cybernetics: Systems, 2018, pp. 1–11.

[8] R. Derakhshani, S.A. Schuckers, L.A. Hornak, L. O'Gorman, Determination of vitality from a non-invasive biomedical measurement for use in fingerprint scanners, Pattern Recogn. 36 (2) (2003) 383–396.

[9] A. Antonelli, R. Cappelli, D. Maio, D. Maltoni, A new approach to fake finger detection based on skin distortion, in: Advances in Biometrics, Springer, 2005, pp. 221–228.

[10] Y. Zhang, J. Tian, X. Chen, Y. Xin, S. Peng, Fake finger detection based on thin-plate spline distortion model, in: International Conference on Advances in Biometrics, Springer-Verlag, 2007, pp. 742–749.

[11] J. Jia, L. Cai, K. Zhang, D. Chen, A new approach to fake finger detection based on skin elasticity analysis, in: Advances in Biometrics, Springer Berlin Heidelberg, 2007, pp. 309–318.

[12] B. Tan, S. Schuckers, New approach for liveness detection in fingerprint scanners based on valley noise analysis, J. Electron. Imaging 17 (2008) 1.

[13] A. Abhyankar, S. Schuckers, Integrating a wavelet based perspiration liveness check with fingerprint recognition, Pattern Recogn. 42 (5) (2008) 452–464.

[14] A. Abhyankar, S. Schuckers, Wavelet-based approach to detecting liveness in fingerprint scanners, in: Proceedings of the SPIE Vol. 5404, Defense and Security Symposium, Biometric Technology for Human Identification, 2004, pp. 278–286.

[15] B. Tan, S. Schuckers, Comparison of ridge—and intensitybased perspiration liveness detection methods in fingerprint scanners, in: Proceedings of SPIE Vol. 6202, Orlando, March 2006.

[16] B. DeCann, B. Tan, S. Schuckers, A novel region based liveness detection approach for fingerprint scanners, in: Advances in Biometrics Lecture Notes in Computer Science, vol. 5558, Springer Berlin Heidelberg, 2009, pp. 627–636.

[17] N. Manivanan, S. Memon, W. Balachandran, Automatic detection of active sweat pores of fingerprint using highpass and correlation filtering, Electron. Lett. 46 (18) (2010) 1268–1269.

[18] S. Nikam, S. Agarwal, Ridgelet-based fake fingerprint detection, Neurocomputing 72 (2008) 2491–2506.

[19] C. Yuan, X. Sun, Fingerprint liveness detection adapted to different fingerprint sensors based on multiscale wavelet transform and rotation-invarient local binary pattern, J. Internet Technol. 19 (1) (2018) 091–098.

[20] W. Kim, Fingerprint liveness detection using local coherence patterns, IEEE Signal Process Lett. 24 (1) (2016) 51–55.

[21] J. Galbally, S. Marcel, J. Fierrez, Image quality assessment for fake biometric detection: application to iris, fingerprint and face recognition, IEEE Trans. Image Process. 23 (2) (2014) 710–724.

[22] A. Abhyankar, S. Schuckers, Fingerprint liveness detection using local ridge frequencies and multiresolution texture analysis techniques, in: IEEE International Conference on Image Processing, 2006, pp. 321–324.

[23] J. Martins, L. Oliveira, R. Sabourin, Combining textural descriptors for forest species recognition, in: Proceedings of the 38th Annual Conference on IEEE Industrial Electronics Society, 2012, pp. 1483–1488.

[24] Z. Xia, C. Yuan, X. Sun, D. Sun, R. Lv, Combining wavelet transform and LBP related features for fingerprint liveness detection, IAENG Int. J. Comput. Sci. 43 (2016) 3.

[25] R.F. Nogueira, R.D.A. Lotufo, R.C. Machado, Evaluating software-based fingerprint liveness detection using convolutional networks and local binary patterns, in: IEEE Workshop on Biometric Measurements & Systems for Security & Medical Applications, IEEE, 2014.

[26] R.F. Nogueira, R.D.A. Lotufo, R.C. Machado, Fingerprint liveness detection using convolutional neural networks, IEEE Trans. Inf. Forensics Secur. 11 (6) (2016) 1206–1213.

[27] C. Yuan, X. Li, Q. Wu, J. Li, X. Sun, Fingerprint liveness detection from different fingerprint materials using convolutional neural network and principal component analysis, Comput. Mater. Contin. 53 (4) (2017) 357–372.

[28] K. He, X. Zhang, S. Ren, J. Sun, Spatial pyramid pooling in deep convolutional networks for visual recognition, IEEE Trans. Pattern Anal. Mach. Intell. 37 (9) (2015) 1904.

[29] A. Krizhevsky, I. Sutskever, G.E. Hinton, ImageNet classification with deep convolutional neural networks, in: International Conference on Neural Information Processing Systems, Curran Associates Inc, 2012, pp. 1097–1105.

[30] Y. Jia, E. Shelhamer, J. Donahue, S. Karayev, J. Long, R. Girshick, S. Guadarrama, T. Darrell, Caffe: convolutional architecture for fast feature embedding, in: ACM International Conference on Multimedia, ACM, 2014, pp. 675–678.

[31] P. Shamsolmoali, M. Zareapoor, J. Yang, Convolutional neural network in network (CNNiN): hyperspectral image classification and dimensionality reduction, Iet Image Processing, April 2018.

[32] L. Ghiani, D. Yambay, V. Mura, S. Tocco, G.L. Marcialis, F. Roli, S. Schuckcrs, Livdet 2013 fingerprint liveness detection competition 2013, in: 2013 International Conference on Biometrics (ICB), IEEE, 2013, pp. 1–6.

[33] C. Gottschlich, E. Marasco, A.Y. Yang, B. Cukic, Fingerprint liveness detection based on histograms of invariant gradients, in: 2014 IEEE International Joint Conference on Biometrics (IJCB), 2014, pp. 1–7.

[34] C. Yuan, Z. Xia, X. Sun, D. Sun, R. Lv, Fingerprint liveness detection using multiscale difference co-occurrence matrix, Opt. Eng. 55 (6) (2016). 063111-1-063111-10.

[35] S. Kim, B. Park, B.S. Song, S. Yang, Deep belief network based statistical feature learning for fingerprint liveness detection, Pattern Recogn. Lett. 77 (2016) 58–65.

[36] P.V. Reddy, A. Kumar, S. Rahman, T.S. Mundra, A new antispoofing approach for biometric devices, IEEE Trans. Biomed. Circuits Syst. 2 (4) (2008) 328–337.

[37] L. Ghiani, G.L. Marcialis, F. Roli, Fingerprint liveness detection by local phase quantization, in: Proc. 21st Int. Conf. Pattern Recognit. (ICPR), November 2012, pp. 537–C540.

[38] J. Jia, L. Cai, Fake Finger Detection Based on Time-Series Fingerprint Image Analysis, the Interpretation of Visual Motion, MIT Press, 2007, pp. 341–345.

[39] L. Ghiani, A. Hadid, G.L. Marcialis, F. Roli, Fingerprint liveness detection using binarized statistical image features, in: IEEE Sixth International Conference on Biometrics: Theory, Applications and Systems, September 2013, pp. 1–6.

[40] D. Gragnaniello, G. Poggi, C. Sansone, L. Verdoliva, Fingerprint liveness detection based on weber local image descriptor, in: 2013 IEEE Workshop on Biometric Measurements and Systems for Security and Medical Applications (BIOMS), IEEE, September 2013, pp. 46–50.

[41] H. Choi, R. Kang, K. Choi, J. Kim, Aliveness detection of fingerprints using multiple static features, in: Proc. of World Academy of Science, Engineering and Technology, 2007, p. 22.

[42] E. Marasco, C. Sansone, Combining perspiration- and morphology-based static features for fingerprint liveness detection, Pattern Recogn. Lett. 33 (9) (2012) 1148–1156.

[43] D. Gragnaniello, G. Poggi, C. Sansone, L. Verdoliva, Local contrast phase descriptor for fingerprint liveness detection, Pattern Recogn. 48 (4) (2015) 1050–C1058.

[44] T. Ojala, M. Pietikainen, T. Maenpaa, Multiresolution gray-scale and rotation invariant texture classification with local binary patterns, IEEE Trans. Pattern Anal. Mach. Intell. 24 (7) (2002) 971–987.

[45] R. Davarzani, K. Yaghmaie, S. Mozaffari, M. Tapak, Copy-move forgery detection using multiresolution local binary patterns, Forensic Sci. Int. 231 (2013) 61–72.

[46] O. Celiktutan, B. Sankur, I. Avcibas, Blind identification of source cell-phone model, IEEE Trans. Inf. Forensics Secur. 3 (3) (2008) 553–566.

[47] S.B. Nikam, S. Agarwal, Texture and wavelet-based spoof fingerprint detection for fingerprint biometric systems, in: First International Conference on Emerging Trends in Engineering and Technology, IEEE Computer Society, 2008, pp. 675–680.

[48] C. Yuan, X. Sun, Q.M.J. Wu, Difference co-occurrence matrix using BP neural network for fingerprint liveness detection, Soft. Comput. 3 (2018) 1–13.

[49] G.L. Marcialis, F. Roli, A. Tidu, Analysis of fingerprint pores for vitality detection, in: 2010 20th International Conference on Pattern Recognition (ICPR), IEEE, 2010, pp. 1289–1292.

[50] S.B. Nikam, S. Agarwal, Local binary pattern and wavelet-based spoof fingerprint detection, Int. J. Biometrics 1 (2) (2008) 141–C159.

[51] R.K. Dubey, J. Goh, V.L.L. Thing, Fingerprint liveness detection from single image using low-level features and shape analysis, IEEE Trans. Inf. Forensics Secur. 11 (7) (2016) 1461–1475.

[52] C. Yuan, Z. Xia, L. Jiang, Y. Cao, Q.M.J. Wu, X. Sun, Fingerprint liveness detection using an improved CNN with image scale equalization, IEEE Access 7 (99) (2019) 26953–26966.

[53] T. Akilan, Q.J. Wu, H. Zhang, Effect of Fusing Features From Multiple DCNN Architectures in Image Classification, vol. 12, IET Image Processing, 2018. no. 7.

[54] S. Parthasaradhi, R. Derakhshani, L. Hornak, S. Schuckers, Time-series detection of perspiration as a liveness test in fingerprint devices, IEEE Trans. Syst. Man Cybern. Part C Appl. Rev. 35 (3) (2005) 335–343.

[55] S. Memon, N. Manivannan, W. Balachandran, Active pore detection for liveness in fingerprint identification system, in: 19th Telecommunications Forum, 2011, pp. 619–622.

[56] C. Yuan, Z. Xia, X. Sun, Q.M.J. Wu, Deep Residual Network with Adaptive Learning Framework for Fingerprint Liveness Detection, IEEE Transactions on Cognitive and Developmental Systems, 2019.

About the authors

Dr. Chengsheng Yuan received his PhD degree from Nanjing University of Information Science and Technology, China, in 2019. From July 2019 till now, he is a research fellow at the Department of Electrical and Computer Engineering at University of Windsor, Canada. At the same time, he is also a Lecturer in the School of Computer and Software, Nanjing University of Information Science & Technology, China. His research interests include biometric recognition, digital forensic, machine learning and deep learning.

Qi Cui received his BS degree from Nanjing University of Information Science and Technology, China, in 2017. He is currently pursuing his PhD degree with Nanjing University of Information Science and Technology, China, in 2017. His research interests include machine learning and deep learning.

Prof. Xingming Sun received his BS degree in Mathematics from Hunan Normal University, China, in 1984, MS degree in Computing Science from the Dalian University of Science and Technology, China, in 1998, and the PhD degree in Computing Science from Fudan University, China, in 2001. He is currently a Professor and the Dean in the College Computer and Software, Nanjing University of Information Science and Technology, China. In 2006, he was a visiting scholar at University College London, UK. From March 2008 to February 2010, he was employed as a visiting professor by University of Warwick in the United Kingdom. He is currently the Dean of the School of Computer and Software, Nanjing University of Information Engineering and the Director of Jiangsu Network Monitoring Engineering Research Center. He is also a Professor in China-USA Computer Research Center, China. He has also served as Member of the expert group of the National Natural Science Foundation of China (NSFC), Deputy Director of Huazhong Evaluation Center in China Information Security Evaluation Center, Member of National Expert Committee on Digital Watermarking and Information Hiding, Member of Specialized Committee on Theoretical Computer of China Computer Society. He is the reviewer of more 10 international journals. He is the general chair of ICCCS (International Conference of Cloud Computing and Security) 2015, 2016, 2017 and 2018. He is a recipient of Science and Technology Progress Award, and a senior member of IEEE. His research interests include network and information security, digital watermarking, and data security in cloud.

Prof. Q.M. Jonathan Wu received his PhD degree in electrical engineering from the University of Wales, Swansea, UK, in 1990. He was affiliated with the National Research Council of Canada for 10 years beginning in 1995, where he became a Senior Research Officer and a Group Leader. He is currently a Professor with in Department of Electrical and Computer Engineering, University of Windsor, Canada. He is a Visiting Professor in the Department of Computer Science and Engineering, Shanghai Jiao Tong University, Shanghai, China. He has published over 300 peer-reviewed papers in computer vision, image processing, intelligent systems, robotics, and integrated microsystems. His current research interests include 3D computer vision, active video object tracking and extraction, interactive multimedia, sensor analysis and fusion, and visual sensor networks. Dr. Wu holds the Tier 1 Canada Research Chair in Automotive Sensors and Information Systems. He is an Associate Editor of the *IEEE Transactions on Neural Networks and Learning Systems, Cognitive Computation*, and the *International Journal of Robotics and Automation*. He was an Associated Editor of the *IEEE*

Transactions on Systems, Man, and Cybernetics—Part A: Systems and Humans. He has served on technical program committees and international advisory committees for several prestigious conferences.

Dr. Sheng Wu is technical officer on Public Technology in World Health Organization, HQ in Geneva, Switzerland. She holds PhD in Science, Technology and Innovation Management, and has extensive experience in leading research on emerging technology at national and international level.

CHAPTER SIX

Protecting personal sensitive data security in the cloud with blockchain

Zhen Yang[a], Yingying Chen[b], Yongfeng Huang[a], and Xing Li[a]
[a]Department of Electronic Engineering, Tsinghua University, Beijing National Research Center for Information Science and Technology, Beijing, China
[b]School of Engineering and Applied Science, University of Virginia, Charlottesville, VA, United States

Contents

Abstract

To protect personal sensitive data in cloud computing environment, certain issues need to be addressed, including data ownership, fine-grained access control, transparency and auditability. While many models have been explored to address these issues, among most of which, some components, such as Cloud Service Provider

(CSP) are required to be trusted. In this chapter, we introduce a trust-free data access model for personal sensitive data protection in the cloud environment. In our model, an access control mechanism is constructed based on the Ethereum blockchain, which requires no trusted party. The smart contract enables fine-grained access control for cloud data based on the blockchain. Data operations including uploading, updating and downloading can be automated processed and logged in our model to ensure transparency and auditability. Comparisons between our model and existing models show that our trust-free model fulfills all requirements on personal sensitive data protection, and brings no extra security risks. Moreover, our model has less burden for data owner from both the computation perspective and communication perspective.

Abbreviations

AES128	Advanced Encryption Standard 128 bits
AI	artificial intelligence
AMD	Advanced Micro Devices, Inc.
API	application programming interface
CPU	central processing unit
CS	cloud server
CSP	cloud service provider
DBE	dynamic broadcast encryption
DO	data owner
DU	data user
EMR	electronic medical record
EVM	Ethereum virtual machine
GB	giga byte
HDFS	Hadoop distributed file system
IoT	near field communication
JAR	Java ARchive
JVM	Java virtual machine
KB	kilo bytes
LSN	local sequence number
MHT	Merkle Hash Tree
P2P network	peer-to-peer network
PBFT	practical Byzantine fault tolerance
PKI	public key infrastructure
PoW	proof of work
RIPEMD160	RACE (Research and development in Advanced Communications technologies in Europe) Integrity Primitives Evaluation Message Digest 160
RAM	random-access memory
SHA256	secure hash algorithm 256 bit
SLA	service-level agreement
SUNDR	secure untrusted data repository
VM	virtual machines

1. Introduction

Recently, Artificial Intelligence (AI) has been widely used in providing service in people's life. To support intelligent service, the service providers need to collect personal data in the process. In cloud computing environment, more and more personal sensitive data are stored on the cloud.

Cloud service offers great convenience for users, by enabling data sharing and online data modification. Compared to other kinds of data, personal sensitive data on the cloud needs to be handled in a much higher level of security. Data owner of the personal sensitive data, wants to both control access and to monitor its data.

For the cloud data access control, data owner always wants to control data access by himself, which means Cloud Service Provider (CSP)'s involvement is not wanted in the process. In order to address this issue, encrypted access control method is created, where data is outsourced to CSP as ciphertext and its access is controlled with key distribution. This method is called encrypted access control. Many algorithms had been explored in encrypted access control area, for example, proxy re-encryption [1] and broadcast encryption [2,3]. Later, attribute-based encryption [4,5] was proposed to reduce the burden of data owner. However, how to revoke users' authorization on fine-grained control access still remains as a difficult problem for these algorithms.

For the auditability, logging mechanism is commonly resorted. Credible logging may face two kinds of attack: (1) tampering attack: it changes the log content, so as to erase any evidence of illegal operation. (2) Collusion attack: the untrusted cloud server may collude with malicious users for interest reasons. To combat tampering attack, many logging models have been invented using digital signature [6] and hash chain structure [7,8]. For collusion attack, automated logging models offer logging independency for the data owner, with the help of programmable JAR (Java ARchive) [9–11]. Although JAR-based automated logging can defend the two attacks, many tools are required to deploy for data owner.

Blockchain technology, which was originally used for cryptocurrency [12,13], offers an alternative solution for preventing log content from being tampered. Recently, blockchain technology becomes increasingly popular being applied to cloud storage service to improve data security.

Since 2014, blockchain-based secure cloud storage service started to appear in research by Storj [14], Sia [15], Filecoin [16], Datacoin [17]. Blockchain is also used to control privacy data access [18]. For sensitive data such as electronic medical records, a blockchain-based record system [19], which can manage data access and permission has been built. Blockchain consensus can ensure the privacy-preserving access control for Internet of Things (IoT) system as well [20]. Besides, cryptographic blockchain model is formalized to build privacy-preserving smart contracts on the blockchain [21]. However, these blockchain-based models still rely on the trust of the cloud service, although many CSP cannot be trusted.

This chapter constructs a data access model for personal sensitive data in the cloud using blockchain and smart contract. This model does not require trust any components within the system. In this model, we propose a blockchain-based access control mechanism, by which data owner can control its cloud data access policy in a fine granularity. Moreover, we establish a cloud data operation protocol based on smart contract on Ethereum blockchain, which fuses blockchain into cloud storage to ensure data transparency and auditability. The advantages of our model include:

1. Our model provides data owner with fine-grained control of its sensitive data in the cloud environment, including authorization granting and revoking.
2. Our model ensures all data uploaded, updated and downloaded to be recorded in log, which cannot be tampered, so as to support auditability.
3. Our model does not increase the burden of data owner. It is easy for CSP to deploy.

The rest of the chapter is organized as follows: Section 2 reviews the related issues in our discussion. Then we introduce the trust-free data protection model architecture in Section 3. In Section 4, we conduct security analysis of our model and present the results of performance experiment. Section 5 summarizes the research contribution of our research. Section 6 concludes the chapter.

2. Issues in discussion

When it comes to personal sensitive data in cloud environment, users have the following expectations on security: first, the users want to monitor how their data is used and want to control whom to be authorized to have access to data. Second, the users' control and access would not depend on another participant (or actor) in the process. Therefore, when designing a model for personal sensitive data protection, the following issues need to be considered:

➤ *Data ownership.* The target model shall guarantee data owner to own and control its personal sensitive data. The data ownership shall be undeniable and unchangeable.

➤ *Fine-grained access control.* Once the access authorization of the cloud data is granted to some users, it is difficult work to change or revoke the authorization. In the target model, the data owner shall be able to grant and revoke the authorization of its personal sensitive data in the cloud with fine granularity.

➤ *Data Transparency and auditability.* In the target model, all data owners shall have complete transparency over their data. Data operation logs, which are transparent to the data owner, authentic and cannot be tampered. They will support auditability.

➤ *Independent of trust upon other actors.* Data security of cloud storage service usually relies on a trusted CSP. However, in reality, CSP is not always trustable. In the target model, data owner shall be able to monitor and control data access without relying on other trusted actors in the system, including CSP.

2.1 Cryptographic methods

In existing research, cryptographic methods have been commonly used for data encryption [22]. Cryptographic methods can support secure remote storage of data [23]. However, it could not support data sharing, therefore, its application was limited.

In order to realize secure data sharing, cloud data shall be accessed by authorized users, whom the data owner trusts. Therefore, some research tasked CSP to conduct data access control. Some key derivation models [24–26] have been proposed as an extension of key management. While there are two important features of cloud data [27], namely *data dynamics* and *sharing group dynamics*, these key derivation models [24–26] could not support *sharing group dynamics*. Later, some cloud data access control mechanisms which could support data confidentiality dynamics were proposed, including proxy re-encryption [1,28–31], broadcast encryption [2,3], and attribute-based encryption [32–36].

There are two conventional ways to improve security level of data on the cloud: preventive measures and deterrent measures [37]. The preventive measures include security protection and privacy enhancing technologies [38]; the deterrent measures use data transparency and auditability [39,40] to pose psychological obstacles for crime. To ensure transparency and audit-ability, data operations logs are commonly used as supporting proof. There are two kinds of log storage methods: distributed log recorded by data itself [9,10,41] and data-centric log recorded by cloud server [42,43].

In 2012, Tan proposed a prototype of data protection using data tracker [41]. Data tracker was a self-executing container, encrypted to prevent data from direct access. With data tracker, data operation log could be recorded automatically, and then log transfer and data leakage detection could be conducted. Sundareswaran proposed a decentralized data accountability framework using the JAR file [9,10]. This framework used JAR to ensure enforced access control and automatic logging. In addition, this work posed an end-to-end confidentiality auditing model, which contained a *push mode* and a *pull mode*. A *push mode* is a type of passive auditing by data owner, while a *pull mode* is a type of proactive auditing when the data owner suspects data leakage. Although the framework is robust against Java Virtual Machine (JVM) compromise attack, it could not support sharing group dynamics.

Ko proposed a framework of data auditing based on data-centric log [37,42,43]. Data-centric log was produced by Flogger through collecting logs of same data from different Virtual Machines (VM). Then visualization and analysis of data-centric logs could perform auditing.

In 2004, Li proposed a fork consistency preserving secure untrusted data storage model Secure Untrusted Data Repository (SUNDR) based on Straw-man file system [6]. Straw-man file system uses corresponding users' key to sign in every time when there is data modifying. Before data modifying, user are always asked to check the signature. In this way, all users, including the data owners, are aware of data confidentiality. However, this model could not prevent signed log of data from being modified, which will effect confidentiality-awareness of the data owner.

In 2009, Itani [7] designed a data storage and processing model with privacy-awareness. In the model, privacy-relevant operations were recorded into logs based on hash chain, thus data owner can be aware of confidenti-ality of its data. The weakness of this model was that only data owner can use and process the data, while data sharing was not supported.

In 2011, Popa [8] designed a secure cloud storage system, CloudProof, which supported Service-Level Agreement (SLA). Dynamic Broadcast Encryption (DBE) was adopted to protect users' authorization with encryp-tion. They design a special log called *attestation*, which is formulated according to designed structure at each time of data reading or writing, and correlated forward-backward with hash chain method. This special log serves as an identifier of a specific user's access of data. Reviewing attestations, data owner can be aware of data confidentiality. However, confidentiality risk still exists when untrusted CSP connects current attestation to a history attestation and discard attestations between them. In 2013, Hwang [44] proposed non-repudiation cloud storage model for the situation of single account on multiple device exchangeable access. The model used Local Sequence Number (LSN) to eliminate the risk of discarding attestations.

2.2 Blockchain methods

Blockchain is the basic underlying technology of Bitcoin [12]. It was first proposed by Satoshi Nakamoto in his construction of a decentralized point-to-point trading system, which requires no trust. Blockchain has been used by Bitcoin, which started to be in operation since 2009. The advantage of blockchain technology is its feature of being decentralized. By using encryption algorithms, time stamps, tree structure, consensus mechanism and reward mechanism, point-to-point transaction based on decentralized credit, are realized in distributed networks. In the networks, the nodes do not need to be trusted. Thus, the disadvantages of conventional centralized system, such as low reliability, low security, low efficiency and high cost, could be overcome.

Blockchain is implemented as a decentralized ledger, which keeps a data structure based on time series chain, as shown in Fig. 1. Each unit of blockchain is called *block*, which includes block header and block body.

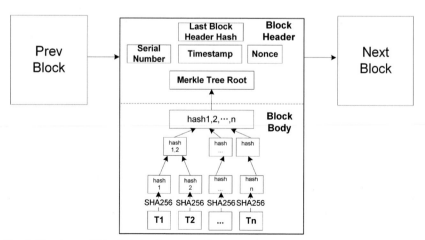

Fig. 1 Structure of blockchain. *Source: author.*

The contents of the block header includes: the hash value of the previous block header, the timestamp, the current Proof of Work (PoW) calculation difficulty value, the solution of the current block PoW problem (the random number that satisfies the requirement), and the Merkle Hash Tree (MHT) root. The block body contains the complete transaction information of a block and is organized in the form of a MHT. All the transactions are aggregated with MHT based on the Secure Hash Algorithm 256 (SHA256) hash function.

The structure of blockchain ensure its security: Except for the creation block, all the other blocks contain the hash value of a previous block header. Meanwhile, by including timestamps, the blockchain is time-series based. The more blocks that are linked behind a certain block, the higher is the cost for modifying that certain block.

As blockchain is built on a Peer to Peer (P2P) network, a consensus mechanism is very important for blockchain's security. In traditional distributed systems, there are many consensus mechanisms, which support strong consistency consensus, for example, PBFT (Practical Byzantine Fault Tolerance), Paxos, Raft. However, the limitations of the scale of these consensus, prevent the P2P network from expanding to large scale network with hundreds of nodes; and they can just tolerant less than $1/3$ of P2P nodes to be malicious. With the development of Bitcoin, a new consistency consensus has been developed, which is called Proof of Work (PoW). The core idea of PoW is to ensure data consistency and consensus security by introducing computational power competition among distributed nodes. In Bitcoin, all the nodes involved in "mining" are traversing to find a random number. This random number makes the result of SHA256 hash function of the block header of the current block less than or equal to a certain value, and finds the random number that meets the requirements. The node obtains the accounting right of the current block and obtains a certain amount of Bitcoin as a reward. In addition, the dynamic difficulty value is introduced so that the time taken to solve the mathematical problem is about 10 min. Thus, tampering a blockchain with PoW consensus is difficult. It would require more than 50% of computation power of the whole blockchain network.

Some researchers extend Blockchain structure to support of peer-to-peer cloud storage service [14,15]. These methods [14,15] could allow the users to transfer and share data without relying on a third party storage provider. Some of them can even support client-side encryption and challenge-response verification.

In 2015, a decentralized personal data management system [18] was proposed. This system used blockchain to remove trust reliance on a third party. This system allowed non-currency transactions, and can store, search and share data.

In 2016, MedRec [19], a decentralized Electronic Medical Record (EMR) management system based on blockchain, was proposed. This system could manage identity authentication, data sharing with confidentiality and accountability. It also motivates users of the system and the medical stakeholders to participate in the network as blockchain "miners."

In 2017, FairAccess [20] was proposed. It was a decentralized pseudonym and privacy-preserving Internet of Things (IoT) authorization management architecture based on blockchain and smart contract. In FairAccess, smart contract is used to express fine-grained and contextual access control policies for decision-making on authorization. Authorization token is utilized to conduct access control.

However, all these exiting blockchain-based models still rely on trust upon CSP, while in cloud environment, in many cases, CSP should not be considered as trustable.

3. Our model: Protecting personal sensitive data security in the cloud with blockchain

In order to address the aforementioned issues, we design a blockchain-based model, which can protect personal sensitive data in the cloud. The terminologies used in our model and their meaning are listed in Table 1:

Our model contains three major components:

➤ Data Owner (DO): DO is the source of personal sensitive data. It has the ownership of the data.

➤ Data User (DU): DU has the demand of accessing DO's personal sensitive data in the cloud.

➤ Cloud Server (CS): CS offers cloud storage service to DO and DU.

In addition to DO, DU and CS, our model also includes a smart contract based on Ethereum blockchain. The structure of our model is shown in Fig. 2. The underlying Ethereum blockchain is a public ledger data structure applied in the Ethereum cryptocurrency system, which is maintained by arbitrary nodes on the Internet with PoW consensus mechanism. PoW consensus and Ethereum network, make Ethereum blockchain untamperable.

In our model, each of the three major components (DO, DU, CS) has an Ethereum account, which holds a pair of keys (public key and private key)

Table 1 Terminologies in our model.

Symbols	Meaning
DO	Data owner
CS	Cloud server
DU	Data user
dbs	The encrypted sensitive data which is divided into many data blocks
dbl_i	The i-th block of the encrypted sensitive data $(i=1,\ldots,n)$
dbh_i	The hash of $\boldsymbol{dbl_i}$ $(i=1,\ldots,n)$
dh	All $\boldsymbol{dbh_i}$ $(i=1,\ldots,n)$
MHT	Merkle Hash Tree
drh	The root hash of the MHT whose leaves are $\boldsymbol{dbh_i}$ $(i=1,\ldots,n)$
dk_{DU}	Decryption key of the sensitive data for DU
pk_{DU}	Public key of DU in Ethereum blockchain system
sk_{DU}	Secret key of DU in Ethereum blockchain system
$[\![dk_{DU}]\!]$	Ciphertext $\boldsymbol{dk_{DU}}$ that is encrypted with $\boldsymbol{pk_{DU}}$

Source: author.

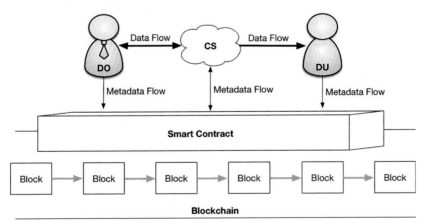

Fig. 2 Our model for personal sensitive data management in cloud environment. *Source: author.*

to support encrypted communication and digital signature. Thus, communication between DO, DU and CS is implemented via Ethereum account transaction mechanism, which guarantees authenticity of communication content with digital signature.

The smart contract based on Ethereum blockchain, offers the feature of recording data access control policies and operation logs into the immutable blockchain structure. The smart contract is designed and deployed by CS to automatically respond to the request from DO/DU. CS can open the contract source code and offer a period of open test for any DO/DU. The contract response first verifies the request identity and authorize permission. Then the contract updates the data access control policy, or records the data operation log after verification.

In our model, data access policy can be granted or revoked by DO in a fine granularity. DO outsources its personal sensitive data in ciphertext form. Therefore, each access control policy includes the data decryption key (dk), which is prepared for the authorized DU. In Fig. 2, updating access control policy is interactions between DO and the smart contract. Because no data operation is involved in access control policy updating, there is no need for CS' participation; this function is triggered by the contract directly. In this way, DO can ensure its data ownership and control the access policy in a fine granularity.

Our model contains a data operation protocol, fused with blockchain-based access policies, to support data transparency and auditability. In this protocol, DO has the right of reading and writing data, whereas DU can only read the data when it is authorized. Our data operation protocol includes both data flow and metadata flow as shown in Fig. 2. For the metadata flow, DO/DU who acts as the initiator, first sends metadata request to the contract. After the contract respond to the request to verify a permission, the metadata of data operation is stored as log in the Ethereum blockchain. For data flow between DO/DU and CS, data operation will be finally executed only after it is confirmed with the log in Ethereum blockchain. In this way, all executed data operations have log records, which cannot be tampered, and the logs are stored in the transparent Ethereum blockchain.

3.1 Identities and pseudonyms

While users and CSP have their own blockchain account, the key pairs and addresses are the unique pseudonym identifier of their identities. The Traditional Public Key Infrastructure (PKI) is centralized, and the PKI management server is not only required to be trustable by all users, but also is tasked to combat various attacks on users' identity in an open network environment. In order to remove these requirements from the cloud server, in our model, we enable all the actors in our system to choose a unique sequence of number as their own private key. Each actor uses a number generator to generate a random stream of numbers; then part of this stream is arbitrarily taken into SHA256 hash calculation.

$$SecretKey = SHA256(RandomStream)$$

Subsequently, we use the elliptic curve cryptographic algorithm defined by the secp256k1 standard [45]. On the elliptic curve field, multiply the *SecretKey* to get the public key of the user.

$$PublicKey = SecretKey*g$$

In the above formula, g is the generator of the elliptic curve domain, and $*$ is the multiplication in the elliptic curve domain. The difficulty of the discrete logarithm calculation, guarantees security of the private key.

In addition, each user can get a unique pseudonym which is derived from RIPEMD160 (RACE (Research and development in Advanced Communications technologies in Europe) Integrity Primitives Evaluation Message Digest 160) hash of its *PublicKey*. Thus every user has a 160-bit pseudonym called *address*.

$$Address = RIPEMD(PublicKey)$$

3.2 Blockchain-based dynamic access control mechanism

Utilizing the *Event Mechanism* in the Ethereum blockchain, we design our data access control policy, as shown in Fig. 3.

In this figure, the structure of Share Event as data access policy consists six parts:

1. The event type as "Share," meaning the access control policy of data sharing.
2. In the second part, "Data User Address" is the authorized DU identifier filled with its Ethereum account address.

Share	Data User Address	Data RootHash	Timestamp	Authority	Datakey Ciphertext encrypted for DU

Fig. 3 Structure of share event as data access policy. *Source: author.*

3. "Data RootHash" (*drh*) field is the only data identifier which is computed by building a MHT for the data blocks.
4. "Timestamp" field is the access policy generation time point.
5. "Authority" field represents the authorization value for DU to the data after this policy is updated: 1 means access authorized and 0 means authorization revoked.
6. The last part is data decryption key distributed to DU: the value is 0, when the value of "Authority" is 0. The last field only has a legal content when "Authority" is 1: dk encrypted by the public key of DU account, which is expressed as $[\![dk_{DU}]\!]$.

The dynamic updating of the access policy enables the dynamic access control sub-protocol, which is shown in Fig. 4. The major steps are listed as below.

1. If DU wants to download data of DO from CS, it first needs to send an Ethereum blockchain transaction to DO to pass *drh* of the requested data.
2. DO can independently decide whether or not to share data with DU after receiving the Ethereum blockchain transaction. If DO would like to share the data, it first calculates dk_{DU} for DU, then encrypts dk_{DU} with the pk_{DU} of DU to generate the $[\![dk_{DU}]\!]$.

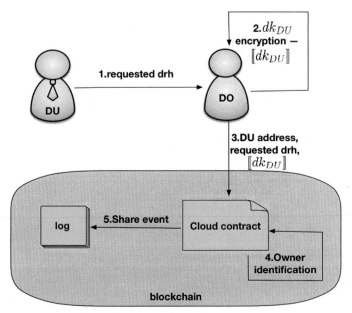

Fig. 4 Protocol of dynamic access policy updating. *Source: author.*

3. DO sends a Ethereum blockchain transaction to the cloud contract passing DU address, *drh*, authority value and $[\![dk_{DU}]\!]$, and assigns the access policy update function of the contract to respond the transaction. When DO wants to revoke DU's access permission for *drh*, DO sends a transaction including DU address, *drh* and revoked authority value to the cloud smart contract. This revocation transaction can also be responded by the access policy updating function of the cloud smart contract.
4. The cloud smart contract verifies if the message sender is DO, who is the only actor that can share data by the contract.
5. If the verification of DO identity pass through, share event is triggered inside the cloud contract, meaning the authorization has been granted to DO to save access policy into the logs on the blockchain. In a share event, DU address, requested *drh* and the $[\![dk_{DU}]\!]$ along with timestamp are recorded.

In the cloud smart contract, its access policy updating function can be triggered by a transaction, and then, the new access policy can be stored into Ethereum blockchain. After receiving a transaction requesting to update access policy, the contract first verifies the validity of the transaction by transaction signature and sender identity. If the transaction is signed and sent by DO of *drh*, the contract will fill in the different information parts (as listed above) of share events and generate the new access control policy event. In this way, a blockchain structure is formulated as the follows (Fig. 5):

Here, except for the information parts of timestamp (T), hash value of the previous block (*Prevhash*), random nonce (N) and transactions (*Txs*), each block includes the hash digest of events ($e_1, \ldots, e_x$) and logs ($l_1, \ldots, l_y$). As shown in Fig. 5, *Prevhash* part makes the relation between adjacent blocks, which can be expressed by the equation:

$$hash\left(T_n, Prevhash_n, N_n, Txs_n, hash(e_1), \ldots, hash(e_x), hash(l_1), \ldots, hash(l_y)\right)$$
$$= Prevhash_{n+1}$$

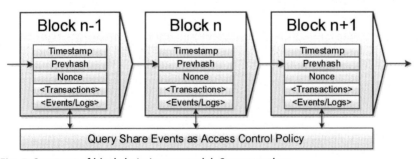

Fig. 5 Structure of blockchain in our model. *Source: author.*

When an event recording access policy is tampered by some malicious nodes, the hash chain between the blocks, including the event and its following block is broken. And the hash chain relation of blockchain is maintained by all the nodes of Ethereum network. Thus, the access control policy in our model is untamperable; and the blockchain-based access policy supports dynamic access control for DO in fine granularity.

3.3 Cloud data operation protocol fused with blockchain-based access policies

When personal sensitive data is outsourced to the cloud, it needs to be encrypted into ciphertext data blocks (dbs); operations of data, including data reading and data have to involve CS. In order to improve security, transparency and auditability level of data operation, we establish in our model a cloud data operation protocol fused with blockchain-based access policies. The data operation protocol not only can control access of cloud data with access policies in the blockchain, but also can ensure all executed operations having their corresponding untamperable logs for transparency and auditability. This data operation protocol is further divided into two sub-protocols, including data writing and data reading.

3.3.1 Data writing sub-protocol

In our model, only DO has permission on data writing, including data uploading and updating. To identify DO's identity, the cloud smart contract maintains a legal data list, which contains all *drh*s and their corresponding DOs' addresses.

Data uploading sub-protocol is shown in Fig. 6. It contains following steps:

1. DO prepares *dbs* and its identifier *drh*.
2. DO sends *dbs* to CS.
3. DO sends a transaction including *drh* to the cloud smart contract and assigns a data uploading function to respond. After the contract receives the transaction, *drh* and its corresponding DO address are saved into the legal data list, which is a temporary array in the cloud smart contract.
4. CS saves received *dbs* into a temporary space. Then CS computes *drh* with *dbs* and sends *drh* to the contract, assigning data uploading confirmation function to respond.
5. The contract compares *drh* from CS with *drh* in its temporary array. If an identical match of *drh* exists, *drh* and its corresponding DO address are taken from the temporary array to save in the legal data list as the contract's state variable. Then, the contract uses blockchain event to

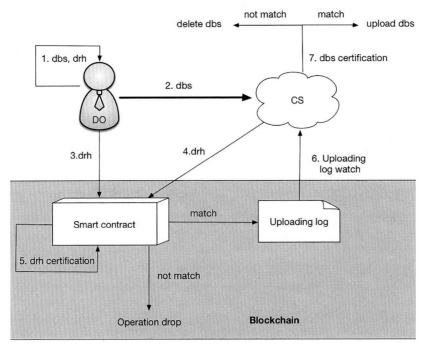

Fig. 6 Sub-protocol of data uploading. *Source: author.*

generate a log into Ethereum blockchain for data uploading operation. Otherwise, the contract terminates the current sub-protocol, if there's no match of *drh*.

6. CS runs a script program to watch new uploading log in the Ethereum blockchain.
7. CS matches *dbs* in the temporary space with the new uploading log. If an identical match of *drh* exists, CS finally saves uploaded *dbs* as legal storage. Otherwise, CS refuses to execute final confirmation and deletes unmatched *dbs*.

The structure of data uploading log is shown in Fig. 7.

In this figure, the structure of upload log for data uploading consists four parts:

1. The log type is defined as data uploading.
2. "Data Owner Address" is the identifier of DO.
3. "Data Roothash" means *drh*, the identifier of the data.
4. The "Timestamp" represents the timepoint of this operation.

The log is generated by the cloud smart contract. The transaction information is filled into corresponding parts.

Upload	Data Owner Address	Data RootHash	Timestamp

Fig. 7 Structure of upload log for data uploading. *Source: author.*

Update	Data Owner Address	new Data RootHash	old Data RootHash	Timestamp

Fig. 8 Structure of update log for data updating. *Source: author.*

In our model, the data updating operation sub-protocol is similar with that of data uploading, except for the following two aspects:

➤ For the data flow, DO only sends part of data (data blocks to be updated) to CS.

➤ For the metadata flow, DO uses both old *drh* and new *drh* to identify the updating operation. In the match step of the cloud contract and CS final confirmation, both of old *drh* and new *drh* should be identically matched. Besides, our data updating log uses the two *drh*s instead of *drh*, as is shown in Fig. 8.

3.3.2 Data reading sub-protocol

Reading data operation in the cloud environment is usually called data downloading. Our data downloading sub-protocol shows a little difference for DO and DU, as shown in Fig. 9.

The data reading sub-protocol contains the following steps.

1. Downloader (DO/DU) sends a transaction including *drh* to the cloud contract and assigns the data downloading function to respond.

2. The contract verifies whether the address of the transaction source account is the same as the address of DO.

3. If the transaction source account is the same as the address of DO, the contract saves DO downloading logs into Ethereum blockchain.

4. If the transaction source is not the same as DO's address, the contract saves Event1 into the blockchain.

5. CS runs a script program to watch Event1 from the blockchain. After a new Event1 is watched, CS queries share event recording *drh* and DU address from the blockchain.

6. If no matching share event exists in the blockchain or the latest entry of matching share events holds an access authority value of 0, CS terminates this protocol.

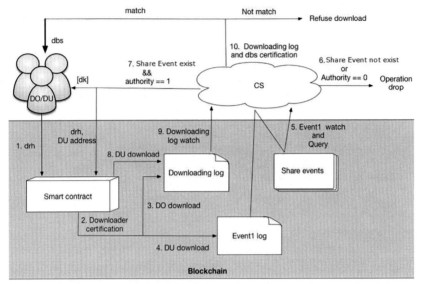

Fig. 9 Sub-protocol of data downloading. *Source: author.*

Down-load	Downloader Address	Data RootHash	Timestamp

Fig. 10 Structure of log for data downloading. *Source: author.*

7. If matching share events exist in the blockchain and the latest entry of them holds an access authority value of 1, CS extracts $[\![dk_{DU}]\!]$ from the access policy and sends it to DU in a transaction. At the same time, CS sends a transaction including *drh* and DU address to the contract to confirm downloading.

8. In the contract, data downloading function is triggered by the confirming downloading transaction from CS. The contract then saves DU downloading log into the blockchain.

9. CS watches new downloading log from the blockchain with the local script program.

10. CS matches the new downloading log with existing cloud data. If the log matches existing *drh*, CS sends the ciphertext data to DU. Otherwise, CS refuses to execute downloading operation.

The structure of log for data downloading is shown in Fig. 10. In the data downloading sub-protocol, the cloud contract fills corresponding information into every field to generate a downloading operation log. Here "download" field defines log type of data downloading. Then the downloader identity, *drh* of the downloaded data and the timepoint of downloading operation is followed.

Finally, DO or authorized DU decrypts downloaded *dbs* to read the plaintext data content.

4. Security analysis and performance evaluation of our model

In this section, we evaluate our model from two main aspects, security proof based on theoretical analysis and performance based on experimental results.

4.1 Security analysis

In order to test the security level of our model, we analyze our model's robustness against possible attack. We also compare our model with the existing models of sensitive data protection to examine its security level.

Although the event logs of Ethereum can be seen by all actors in the system, the malicious users cannot use the information of the operation logs, which only includes DO/DU address, *drh*, $[\![dk_{DU}]\!]$ and timestamp. These information have no potential for leaking owner's privacy, due to the following reasons:

1. One's identity cannot be deduced from its address;
2. The plaintext of the data cannot be inferred from its *drh*;
3. Decrypting $[\![dk_{DU}]\!]$ requires the sk_{DU}, so data content cannot be decrypted by unauthorized users;
4. No essential loss can be caused by published timestamp.

Furthermore, an attacker who wants to modify part of a log entry or to delete a whole log entry can hardly succeed, due to PoW consensus mechanism of the Ethereum blockchain. In order to tamper the blockchain, the attacker has to at least modify the data of the majority of the nodes in the blockchain, which is nearly impossible.

4.1.1 Data operation protocol attack

The attack on data operation protocol can be mainly divided into three types.

4.1.1.1 Untrusted CS in data writing

The tasks of CS in data writing are calculating *drh* and matching new uploading log with received *dbs*. If CS conducts untrusted behavior in either of the two tasks, our data operation protocol will output an error and terminate this data operation. DO then finds out the current service provider of CS un-trustable and changes to another cloud service provider. Data (*dbs*) has no risk of being tampered or leaking.

4.1.1.2 Untrusted CS in data reading

The tasks of CS in data reading include watching Event1, querying share events and matching new downloading log with existing *dbs*. Once CS cheats in these steps for authorized DU downloading, DU can inform DO that CS is untrusted and DO will change cloud service. When CS cheats in these steps for unauthorized DU downloading, the unauthorized DU can only receive *dbs* without $[\![dk_{DU}]\!]$ and the downloading log which has no matching share event in the blockchain, will reveal the untrusted behavior of CS. Data (*dbs*) leakage is prevented in this operation.

4.1.1.3 Malicious DU in data reading

An unauthorized/revoked DU cannot disguise to be authorized, because transaction signature is requested. So malicious DU could only collude with untrusted CS to acquire *dbs*. As elaborated in 2, malicious DU cannot obtain $[\![dk_{DU}]\!]$ to decrypt *dbs*.

4.1.2 Cloud smart contract attack

Since the operation logs are generated during the contract execution, the security of smart contract is the center of the security of our model. The possible attack to the smart contract may take place as the follows:

4.1.2.1 Malicious contract attack

Malicious CS may use a malicious smart contract to harm the interest of users in two aspects:
➢ Recording incorrect owner information in the smart contract;
➢ Generating incorrect operation logs or not generating logs.
There are two scenarios where CS can conduct attack:
1. Deploy an evil contract when cloud service provider starts the cloud storage service for sensitive data;
2. After publishing the cloud storage service and gaining trust from users, CS secretly alters the code of the smart contract.
For the first scenarios, as mentioned, in our model, the code of the smart contract is open source and the users are able to test the smart contract before use it. Therefore, the users can reject to use the cloud storage

service with a suspicious smart contract. For the second scenario, since the code of the contract is stored on the blockchain, CS has to alter the code of the contract by over 50% of the blockchain network computation power, which is nearly impossible to achieve.

4.1.2.2 Fake contract attack

Attackers attempting to use a fake contract as the legal one can hardly succeed. There is an essential characteristic of smart contracts that can combat this kind of attack: after being deployed, each smart contract owns a unique contract address. The contact address is a 160-bit hash calculated by a cryptography hash function, and it called RIPEMD-160.

There are two scenarios of fake contract attack (categorized by who is the attacker):

1. If the attacker is not CS, it has to break through two defense lines in our model: (1) To counterfeit a contract account with the exactly same address as that of the legal one, it needs to keep making collision attack to RIPEMD-160 until it gets the same hash address. (2) Since there will be a condition similar to the Eclipse Attacks on Bitcoin's Peer-to-Peer Network if two contracts with the same address are deployed on the blockchain, the attacker has to control over 50% of blockchain network computation power to make the fake contract work.
2. If the attacker is CS, it must try all ways to get a same address as that of the legal contract address by making collision attack to RIPEMD-160.

Due to the security of RIPEMD-160 hash function, success of a collision attack is nearly impossible. In addition, it's unrealistic for an attacker to control over 50% of blockchain network computation power. Therefore, our model is robust against attack in both these two scenarios.

4.1.2.3 Compromised EVM attack

Since the Ethereum Virtual Machine (EVM) is the runtime environment for the cloud smart contract, attackers may compromise the EVM in order to stop or manipulate the contract execution. However, EVM focuses on preventing denial-of-service attack, which has become somewhat common in the cryptocurrency. To conduct a compromised EVM attack, the attacker needs to compromise more than half of the Ethereum blockchain network computation power, which is nearly impossible. Therefore, our model is robust against compromising EVM attack.

4.1.3 Man-in-the-middle attack

An attacker may intrude communication in our model. We go through below all communication phases in our model, to examine their security status.

4.1.3.1 DO-CS/CS-DU

Only data flow is involved in this phase. In our model, DO does not totally trust CS, so rather than outsourcing the plaintext of the data, it encrypts the data and only outsources the ciphertext of the data: *dbs* when uploading or *dbl$_i$* when updating. Therefore, no matter during uploading or updating process from DO to CS or downloading process from CS to DU, the data is encrypted. The attacker cannot decrypt data, because he does not have the decryption key.

4.1.3.2 DU-DO/DO-contract/DU-contract

The communication between them are metadata flow, which transits between Ethereum accounts. An attacker may intrude a transaction and alter the data of the transaction. For example, when DO sends a transaction containing its address and *drh* to the smart contract to record an uploading operation, an attacker can break into the transaction and alter DO's address to its own address, so that he becomes the "owner" of the data. In our model, the transactions are signed by their senders. So if the transaction is tampered, the Ethereum blockchain miners will not accept it, because the signature of the message cannot pass through certification.

4.1.4 Performance comparison with the existing methods

We compare our model with two existing methods from four security perspectives. The results are shown in Table 2.

All of the three methods have powerful data ownership with DO signature verification. In Zyskind's Blockchain model [18] and our model, fine-grained access control is implemented by updating authority state in the blockchain, supporting both authority granting and revoking. Yang's JAR model [11] does not support fine-grained revocation. All three models can ensure transparency and auditability by logging. Last but not least, no trusted party is required by automated JAR program [11] and by smart contract in our model, which is called trust-free property. Zyskind's Blockchain model [18] needs a trusted cloud server. Overall, our model has better security level in the four aspects than the other two methods [11,18].

Table 2 Security comparison with two existing methods.

Security property	Yang's JAR model [11]	Zyskind's Blockchain model [18]	Our trust-free data protecting model
Data ownership	√	√	√
Fine-grained access control	×	√	√
Transparency and auditability	√	√	√
Trust-free	√	×	√

Source: author.

4.2 Experiment on performance of our model

In this sector, we first introduce the setting of our experiment, then present the results.

4.2.1 Experimental settings

We use two computers with Low configuration, 1 Intel Core i5-2430M CPU (Central Processing Unit) and 4GB (Giga Bytes) RAM (Random-Access Memory) to work as DO and DU. Both of them have comparatively low computation capacity. CS is a powerful server with 4 AMD (Advanced Micro Devices, Inc.) Opteron 8380 CPUs and 64GB RAM. Both DO and DU are installed with the Ethereum client parity to enable transmission of ciphertext message encrypted with accounts' public key, whereas CS has the Ethereum client Geth to conveniently execute JavaScript, which supports CS to watch and query events.

We use AES128 (Advanced Encryption Standard 128) algorithm for DO to encrypt its personal sensitive data. Thus, dk_{DU} size is 128 bits. We use solidity to establish the smart contract for CS, and then deploy the cloud contract on the Ropsten testnet of Ethereum. The blockchain of Ropsten testnet uses a consensus of PoW. The cloud contract will make four different kinds of event logs on operations, including data uploading, updating, sharing and downloading.

Because event logs are not accessible in the contract defined by solidity, we resort to the Web3.js API (Application Programming Interface) of Ethereum to interact with the cloud smart contract. CS executes a set of monitor program to watch the current event log and to retrieve the log history. In addition, a monitor program is running on the Geth client of CS, so that it can always respond to the events of the cloud contract to support our data operation protocol.

We use Web3.js API for DO to build and to update MHT. When DO uploads or updates data in CS, it will resort to its MHT building algorithm to obtain new *drh* of the data. Moreover, for both DO and DU, operations on data can be done by sending transaction to the cloud contract with Web3.js.

4.2.2 Evaluation procedure

We examine the performance of our model from the following aspects:

- ➢ Essential extra cost for DO/DU under three different conditions, during data encryption and MHT building. The three conditions are authorization and revocation, data writing operation and data reading operation.
- ➢ Log retrieval efficiency, operation log size and log watching time cost for CS.

From the perspectives of both computation and communication burden of DO and DU, we conclude that our model offers lightweight personal sensitive data protection service for all cloud users and acceptable cost for cloud service providers. As the Zyskind's Blockchain model [18] doesn't offer a code implementation, we could only compare the performance of our model with the Yang's JAR model [11].

We use the Web3.js API to measure the time cost of DO, and obtain the time cost of CS by comparing the timestamp within the monitor program with the timestamp of a responding log. The results show that our model only increases a little cost for users to protect personal sensitive data in the cloud.

Based on the data operation sub-protocols, we test the conditions mentioned above. In our experiments, the uploaded data has 32 data blocks with a volume size of 2 Giga Bytes, while the updated data has 16 data blocks of them whose volume size is 1 Giga Bytes. Here we use the data block size of Hadoop Distributed File System (HDFS), which is the file system of the most popular big data platform Hadoop.

As our model includes the Ethereum blockchain, the time cost of many processes relies on the blockchain network condition. Through over 100 groups of tests, we find that they are often stable in a range. Here we give observation time of three processes.

- ➢ $T(SQ)$ is the time of share events query executed by CS in its geth client. When we query in 35000 blocks in October 30, 2017, $T(SQ) \in [6.6, 7.1]$ *s*. As the Ropsten testnet of Ethereum grows 5600 *blocks/day* and $T(SQ)$ increases 0.05 *ms/block*, $T(SQ)$ grows 0.28*s/day*.

> $T(LW)$ is watching a new log since it's generated in the blockchain. We measure CS reacting time to a latest event log by subtracting the timestamp of the watching function with the log's timestamp. In our model, it usually takes $[0,30]$ seconds for CS to watch each new event log with an average cost of $11.7s$.

> $T(Tx)$ is the time cost of sending a transaction. We compare the sending time of a transaction with the transaction's timestamp to acquire the time cost of a transaction. This time interval makes sense because the timestamp of the transaction is the end time of contract execution. As the cloud smart contract is executed in EVM which is a distributed network, transaction cost time is often in a range $[5,25]$ seconds with an average cost of $15.25s$.

4.2.2.1 Authorization and revocation phase

We use two accounts for DU and DO. DU is a user which DO would like to share data with. DU sends a transaction containing drh to DO. After DO receives the transaction, it encrypts the dk_{DU} with the pk_{DU} to get $[\![dk_{DU}]\!]$ and then send transaction to the cloud smart contract passing DU address, drh and the $[\![dk_{DU}]\!]$ to its share function. We can see from the result that the share event watching information printed on the console and find the sharing record of our test from sharing logs. As we use CP-ABE (Ciphertext Policy-Attribute-based Encryption) [5] algorithm to encrypt our data, the dk_{DU} in our model is 64 bytes. When we use the parity module to encrypt dk_{DU} with pk_{DU}, we find that $[\![dk_{DU}]\!]$ is 177 bytes. In this scenario, we conduct test for more than 10 times.

In Table 3, we compare Yang's JAR model with our model on access authorization. Revocation in our model has the same performance with authorization, whereas Yang's JAR model [11] does not support fine-grained revocation. In Yang's JAR model [11], computation cost of generating new Client JAR is $0.08s$, while generating $[\![dk_{DU}]\!]$ costs only $0.012s$.

Table 3 DO Burden in authorization.

	Yang's JAR model [11]	Our trust-free data protecting model
Computation	Generate new Client JAR	Generate $[\![dk_{DU}]\!]$
Communication	Updating sharing group list message (to CS) and Client JAR (to DU)	Tx with drh and $[\![dk_{DU}]\!]$

Source: author.

For communication cost, updating sharing group list message costs 177 $bytes = 0.173$ KB(Kilo Bytes), the size of Client JAR is 13.2 KB and the size of transaction with drh and $[\![dk_{DU}]\!]$ is no more than 0.3 KB. Therefore, our model has better performance in authorization.

4.2.2.2 Data writing

Firstly, DO encrypts a plaintext data to get dbs and then calculate its dh. Secondly, DO builds the MHT for dh to generate drh. Finally, DO sends drh to the cloud contract. As for the result, we can see the uploading log watching information printed on the console and also find the uploading record of our test from uploading logs. In this scenario, we have conducted test for more than 50 times. The main time cost of DO relies on the efficiency of data encryption and the MHT building which is positively related to the size of the uploaded data. The encryption tests is closely related with the encryption algorithm, while the MHT building tests cover a range from 2 to 24,576 for the size of drh. The MHT building test is shown in Fig. 11.

For DO: Using the data block size of HDFS, the MHT building time cost of 2GB data is 1.56 s as shown in Fig. 11. As Fig. 11 shows, it is reasonable that building MHT time cost increases linearly as the data size increases.

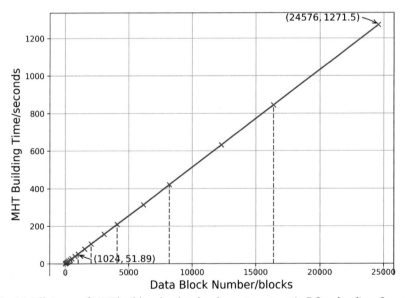

Fig. 11 Efficiency of MHT building by the cloud smart contract in DO uploading. *Source: author.*

The time cost by DO in our model is acceptable to some extent, as we assume that DO of sensitive data would like to spend some time for better data security.

For CS: the log of uploading stores three parts of information, including the hash of the signature of the event, DO address, *drh*, and a timestamp in its data area. Since each part of the information (such as the timestamp) costs only 32 bytes storage space, the time cost of the logging is very small.

DO first encrypts the new partially data with CP-ABE, and then calculates their data hashes. Then DO updates *drh* based on scenario 1 with these new data hashes in indicated positions and builds a new MHT to calculate new *drh*. Finally, DO sends a transaction to the cloud contract passing old *drh* and new *drh* to its update function. We can see from the result that the update log watching information printed on the console and find the updating record of our test from updating logs.

In this scenario, we conduct test for more than 50 times. The main tasks of DO are data encryption, updating old *drh* and building new MHT. Data encryption and MHT building here are in the same way as in the uploading sub-protocol, so they are not analyzed again here. The cost of updating *drh* is depended on the number of new *dbh*. Our tests cover a size range for new *dbh* from 2 to 16,384, as is shown in Fig. 12.

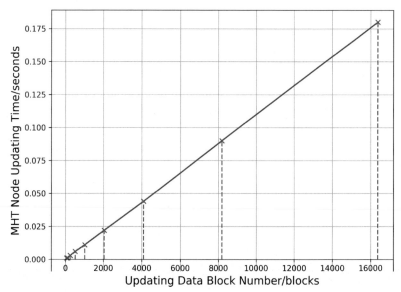

Fig. 12 MHT node updating efficiency by the cloud smart contract in DO updating. *Source: author.*

For DO: To prepare the new data blocks, for the MHT node updating, it is certain the new *dh* generating time is positively related to the new *dbh* number. The time cost of updating *dh* is so small that the time cost of MHT updating is almost the same as the time cost of MHT uploading.

For CS: the log of update store 5 parts of information, including the hash of the signature of the event, DO address, old *drh*, new *drh*, and a timestamp in its data area. Comparing to the uploading logs, it only increases one part of information, which is 32 bytes, and the time cost for the logging is still very tiny.

From Table 4, we can see the differences of DO burden between our model and Yang's JAR model [11]. In data uploading, our model has lower computation burden than Yang's JAR model [11]. As less data blocks are encrypted in data updating in our model, updating data costs less burden on DO, while Yang's JAR model [11] does not support data updating.

For the communication cost of DO, our model has different metadata from that of Yang's JAR model [11], as shown in Table 4. Fig. 13 is the comparison of communication cost between two models. It can be seen that our model achieves lower communication burden for DO in uploading. DO updating communication burden has much smaller data ciphertext and nearly the same size of transaction, which means data updating communication cost for DO is smaller than data uploading.

Table 4 DO burden in data writing operation.

	Yang's JAR model [11]	Our trust-free data protecting model upload	Our trust-free data protecting model update
Computation	Data blocks encryption	Data blocks encryption	Partial data blocks encryption
	Compute all blockhashes and file *drh*	Compute all blockhashes and file *drh*	Compute partial blockhashes and new file *drh*
	Generate JAR		
Communication	Ciphertext data	Ciphertext data	Partial ciphertext data blocks
	Server JAR	Tx with *drh*	Tx with old *drh* and new *drh*

Source: author.

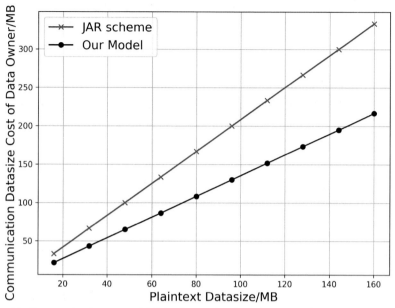

Fig. 13 Communication cost of DO in data uploading. *Source: author.*

In these scenarios, CS computation cost is $T(LW) + T(Tx)$. It is an afford-able cost for CS and acceptable for DO.

4.2.2.3 Data reading

In Table 5, we compare Yang's JAR model with our model in data down-loading operation. We can see that the two models have the same cost on DO downloading and DU downloading. Fig. 14 shows that DO burden of our model in computation is much smaller. In communication, the permis-sion size is 0.327 *KB* while the transaction with $\llbracket dk_{DU} \rrbracket$ is no more than 0.2 *KB*. Therefore, our model has lower burden on both DO and DU in downloading data.

4.2.2.4 Cloud server's computation and communication cost

We present the computation and communication burden of CS in different operation in Table 6. CS computation cost differs between DO down-loading and DU downloading, while CS communication takes the same time for sending ciphertext data *dbs* in both situation.

For computation cost, we show in Fig. 15 the latency level of different operation. CS computation cost of DO downloading is $T(LW)$, while the cost of DU downloading is $2T(LW) + T(SQ) + T(Tx)$. While $T(LW)$,

Table 5 DO burden in downloading.

	Yang's JAR model [11]	Our trust-free data protecting model
Computation	Verify permission	$[\![dk_{DU}]\!]$ Decryption
	Data decryption	Data decryption
Communication	Permission message	Tx with $[\![dk_{DU}]\!]$
	Ciphertext data	Ciphertext data

Source: author.

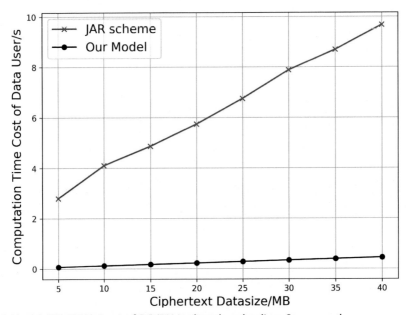

Fig. 14 Computation cost of DO/DU in data downloading. *Source: author.*

Table 6 CS burden in different operation.

CS burden	Authorization and revocation	Data writing	Data reading of DO	Data reading of DU
Computation	$T(LW)$	$T(LW)$ $+ T(Tx)$	$T(LW)$	$2T(LW)$ $+ T(SQ)$ $+ T(Tx)$
Communication		Ciphertext data	Ciphertext data	Ciphertext data

Source: author.

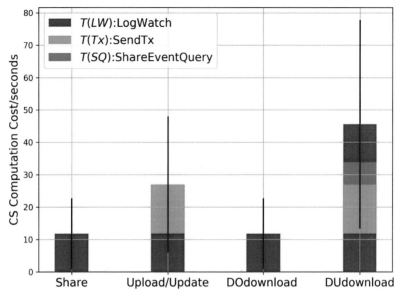

Fig. 15 Computation cost in CS for different operations. *Source: author.*

$T(SQ)$ and $T(Tx)$ are floating with the Ethereum network condition, we test latency of different operations for 100 times to obtain the latency average level and varying range. From Fig. 15, CS burden still keeps an affordable latency for CS and acceptable for DO/DU.

5. Summary of contributions of the research

Our model makes the following innovative contributions to the research on data security in the cloud environment:

➢ Our model provides the data owner with fine-grained control ownership of its sensitive data in the cloud, including authorization granting and revoking.

➢ Our model ensures every executed cloud data uploading, updating and downloading operation to be recorded on an untamperable log, which supports auditability.

➢ Our model does not increase much burden for the data owner, and it is easy for CSP to deploy.

Despite of the above advantages of our model, some problems need to be resolved in future research, for example:

➢ Due to the blockchain transaction delay, the data operation delay becomes larger.

➢ The blockchain-based access policy query time increases along with the size of the blockchain.

On our research agenda, we will explore the mechanism which can stabilize and minimize the transaction delay in blockchain in the cloud environment; we will also explore method for optimizing access policy query that can cope with the burden brought by the increasing size of blockchain.

6. Conclusion

The data owner does not want its personal sensitive data falling in the hand of any other entities. Meeting this security need is challenging when the data is outsourced in the cloud. We employ Ethereum blockchain and smart contract, which include a blockchain-based access control mechanism and a cloud data operation protocol fused with blockchain-based access policies, to construct a model that can protect personal sensitive data security in the cloud. Our model offers data owner a light-burdening solution, which enables fine-grained access control and data transparency without trust dependence on CSP. The experimental results show that our model having better security level and adding less burden to cloud users (both DO and DU), comparing with other existing models.

Acknowledgments

The authors would like to thank the anonymous referees for their valuable comments and helpful suggestions. The work is supported by the National Key Research and Development Program of China (No. 2016YFB0800402) and the National Natural Science Foundation of China (No. U1836204, No. U1536207) and China Postdoctoral Science Foundation (No. 2018M641356).

Key terminology and definitions

Cloud Service Provider (CSP) A CSP is a company that delivers cloud computing based services and solutions to businesses and/or individuals. This service organization may provide rented and provider-managed virtual hardware, software, infrastructure and other related services. Cloud services are becoming increasingly desirable for companies because they offer advantages in terms of cost, scalability and accessibility. A cloud provider is also known as a utility computing provider. This role is typically related to that of a managed service provider (MSP), but usually, the latter provides other managed IT solutions.

Java Archive (JAR) A JAR (Java ARchive) is a package file format typically used to aggregate many Java class files, their associated metadata and resources (text, images, etc.) into one file for distribution.

Virtual Machine (VM) A VM is an emulation of a computer system. Virtual machines are based on computer architectures and provide functionality of a physical computer. Their implementation may involve specialized hardware, software, or a combination.

Dynamic Broadcast Encryption (DBE) Broadcast encryption is the cryptographic problem of delivering encrypted content (e.g., TV programs or data on DVDs) over a broadcast channel in such a way that only qualified users (e.g., subscribers who have paid their fees or DVD players conforming to a specification) can decrypt the content. The challenge arises from the requirement that the set of qualified users can change in each broadcast emission, and therefore revocation of individual users or user groups should be possible using broadcast transmissions, only, and without affecting any remaining users. As efficient revocation is the primary objective of broadcast encryption, solutions are also referred as revocation schemes.

Merkle Hash Tree (MHT) In cryptography, a hash tree or Merkle tree is a tree-shape structure, where each leaf node is labeled with the hash of certain data block, and each node is labeled with the cryptographic hash of the labels of its child nodes. Hash trees allow efficient and secure verification of the content of large data structures. A Merkle tree is recursively defined as a binary tree of hash lists, where the parent node is the hash of its children, and the leaf nodes are hashes of the original data blocks.

Peer–to–Peer (P2P) Network Peer-to-peer (P2P) computing or networking is a distributed application architecture that partitions tasks or workloads between peers. Peers are equally privileged, equipotent participants in the application.

Secure Hash Algorithm 256 (SHA256) SHA256 stands for Secure Hash Algorithm—256 bit and is a type of hash function commonly used in blockchain. A hash function is a type of mathematical function. It turns data into its fingerprint, which is called hash. SHA256 is designed by the United States National Security Agency (NSA) using the Merkle–Damgård structure, from a one-way compression function, which is based on the Davies–Meyer structure.

Proof of Work (PoW) A PoW system (or protocol, or function) is an economic measure to deter denial of service attacks and other service abuse, such as spam on a network, by requiring some work from the service requester, usually meaning processing time by a computer. The concept was invented by Cynthia Dwork and Moni Naor as presented in a journal article in 1993. The term "Proof of Work" or PoW was first coined and formalized in a 1999 paper by Markus Jakobsson and Ari Juels.

References

[1] G. Ateniese, K. Fu, M. Green, S. Hohenberger, Improved proxy re-encryption schemes with applications to secure distributed storage, ACM Trans. Inf. Syst. Secur. 9 (1) (2006) 1–30.
[2] S. Chow, C.K. Chu, X. Huang, J. Zhou, R.H. Deng, Dynamic Secure Cloud Storage with Provenance, Cryptography and Security: From Theory to Applications, Springer, Berlin, Heidelberg, 2012, pp. 442–464.
[3] K.-L. Tsai, J.-S. Tan, F.-Y. Leu, Y.-L. Huang, A group file encryption method using dynamic system environment key, in: 2014 17th International Conference on Network-Based Information Systems (NBiS), IEEE, 2014, pp. 476–483.
[4] A. Sahai, B. Waters, Fuzzy identity-based encryption, in: Annual International Conference on the Theory and Applications of Cryptographic Techniques, Springer, Berlin, Heidelberg, 2005, pp. 457–473.

[5] J. Bethencourt, A. Sahai, B. Waters, Ciphertext-policy attribute-based encryption, in: 2007 IEEE Symposium on Security and Privacy(SP'07), IEEE, 2007, pp. 321–334.

[6] J. Li, M.N. Krohn, D. Mazieres, D. Shasha, Secure untrusted data repository (SUNDR), in: The 6th Symposium on Operating Systems Design and Implementation(OSDI'04), USENIX, 2004, pp. 121–136.

[7] W. Itani, A. Kayssi, A. Chehab, Privacy as a service: privacy-aware data storage and processing in cloud computing architectures, in: The 8th IEEE International Conference on Dependable, Autonomic and Secure Computing(DASC'09), IEEE, 2009, pp. 711–716.

[8] R.A. Popa, J.R. Lorch, D. Molnar, H.J. Wang, L. Zhuang, Enabling security in cloud storage SLAs with CloudProof, in: USENIX Annual Technical Conference, vol. 242, USENIX, 2011, pp. 355–368.

[9] S. Sundareswaran, A. Squicciarini, D. Lin, S. Huang, Promoting distributed accountability in the cloud, in: 2011 IEEE International Conference on Cloud Computing (CLOUD), IEEE, 2011, pp. 113–120.

[10] S. Sundareswaran, A. Squicciarini, D. Lin, Ensuring distributed accountability for data sharing in the cloud, IEEE Trans. Dependable Secure Comput. 9 (4) (2012) 556–568.

[11] Z. Yang, W. Wang, Y. Huang, Ensuring reliable logging for data accountability in untrusted cloud storage, in: 2017 IEEE International Conference on Communications (ICC), IEEE, 2017, pp. 1966–1971.

[12] S. Nakamoto, Bitcoin: A Peer-to-Peer Electronic Cash System, 2008. https://bitcoin.org/bitcoin.pdf.

[13] V. Buterin, Ethereum: A Next-Generation Smart Contract and Decentralized Application Platform, 2014. https://ethereum.org/en/whitepaper/.

[14] S. Wilkinson, T. Boshevski, J. Brandoff, V. Buterin, Storj A Peer-to-Peer Cloud Storage Network, 2014. https://storj.io/storj2014.pdf.

[15] D. Vorick, L. Champine, Sia: Simple Decentralized Storage, 2014. https://sia.tech/sia.pdf.

[16] Protocol Labs, Filecoin: A Decentralized Storage Network, 2017. https://filecoin.io/filecoin.pdf.

[17] Datacoin, Technical Details, 2013. http://datacoin.info/technical-details/.

[18] G. Zyskind, O. Nathan, Decentralizing privacy: using blockchain to protect personal data, in: 2015 IEEE Security and Privacy Workshops (SPW), IEEE, 2015, pp. 180–184.

[19] A. Azaria, A. Ekblaw, T. Vieira, A. Lippman, Medrec: using blockchain for medical data access and permission management, in: International Conference on Open and Big Data (OBD), IEEE, 2016, pp. 25–30.

[20] A. Ouaddah, A.A. Elkalam, A.A. Ouahman, Towards a novel privacy-preserving access control model based on blockchain technology in IoT, in: Europe and MENA Cooperation Advances in Information and Communication Technologies, Springer, Cham, 2017, pp. 523–533.

[21] A. Kosba, A. Miller, E. Shi, Z. Wen, C. Papamanthou, Hawk: the blockchain model of cryptography and privacy-preserving smart contracts, in: 2016 IEEE Symposium on Security and Privacy(SP), IEEE, 2016, pp. 839–858.

[22] D. Chen, H. Zhao, Data security and privacy protection issues in cloud computing, in: 2012 International Conference on Computer Science and Electronics Engineering (ICCSEE), IEEE, 2012, pp. 647–651.

[23] M. Blaze, A cryptographic file system for UNIX, in: Proceedings of the 1st ACM Conference on Computer and Communications Security, ACM, 1993, pp. 9–16.

[24] E.-J. Goh, H. Shacham, N. Modadugu, D. Boneh, SiRiUS: securing remote untrusted storage, in: The Network and Distributed System Security Symposium (NDSS'03), The Internet Society, 2003, pp. 131–145.

[25] S.D.C. Di Vimercati, S. Foresti, S. Jajodia, S. Paraboschi, P. Samarati, Over-encryption: management of access control evolution on outsourced data, in: Proceedings of the 33rd International Conference on Very Large Data Bases(VLDB), VLDB, 2007, pp. 123–134.

[26] M. Kallahalla, E. Riedel, R. Swaminathan, Q. Wang, K. Fu, Plutus: scalable secure file sharing on untrusted storage, in: The 2nd USENIX Conference on File and Storage Technologies(FAST), USENIX, 2003, pp. 29–42.

[27] B. Wang, H. Li, M. Li, Privacy-preserving public auditing for shared cloud data supporting group dynamics, in: The 2013 IEEE International Conference on Communications (ICC), IEEE, 2013, pp. 1946–1950.

[28] W. Wang, Z. Li, R. Owens, B. Bhargava, Secure and efficient access to outsourced data, in: Proceedings of the 2009 ACM Workshop on Cloud Computing Security, ACM, 2009, pp. 55–66.

[29] S. Kamara, K. Lauter, Cryptographic cloud storage, in: International Conference on Financial Cryptography and Data Security, Springer, Berlin, Heidelberg, 2010, pp. 136–149.

[30] K. Liang, J.K. Liu, D.S. Wong, W. Susilo, An efficient cloud-based revocable identity-based proxy re-encryption model for public clouds data sharing, in: European Symposium on Research in Computer Security, Springer, Berlin, Heidelberg, 2014, pp. 257–272.

[31] J. Baek, Q.H. Vu, J.K. Liu, X. Huang, Y. Xiang, A secure cloud computing based framework for big data information management of smart grid, IEEE Trans. Cloud Comput. 3 (2) (2014) 233–244.

[32] S. Yu, C. Wang, K. Ren, W. Lou, Achieving secure, scalable, and fine-grained data access control in cloud computing, in: The 2010 IEEE Conference on Computer Communications (INFOCOM'10), IEEE, 2010, pp. 1–9.

[33] J. Liu, X. Huang, J.K. Liu, Secure sharing of personal health records in cloud computing: ciphertext-policy attribute-based signcryption, Futur. Gener. Comput. Syst. 52 (2015) 67–76.

[34] S. Tu, Y. Huang, Towards efficient and secure access control system for mobile cloud computing, China Commun. 12 (12) (2015) 43–52.

[35] S. Niu, S. Tu, Y. Huang, An effective and secure access control system model in the cloud, Chin. J. Electron. 24 (3) (2015) 524–528.

[36] S. Tu, S. Niu, H. Li, A fine-grained access control and revocation model on clouds, Concurr. Comput. 28 (6) (2016) 1697–1714.

[37] R.K. Ko, Data Accountability in Cloud Systems, Security, Privacy and Trust in Cloud Systems, Springer, Berlin, Heidelberg, 2014, pp. 211–238.

[38] G. Danezis, J. Domingo-Ferrer, M. Hansen, J.-H. Hoepman, D.L. Metayer, R. Tirtea, et al., Privacy and Data Protection by Design-From Policy to Engineering, arXiv preprint arXiv:1501.03726, European Union Agency for Network and Information Security, 2015.

[39] S. Pearson, V. Tountopoulos, D. Catteddu, M. Südholt, R. Molva, C. Reich, et al., Accountability for cloud and other future internet services, in: 2012 IEEE 4th International Conference on Cloud Computing Technology and Science (CloudCom), IEEE, 2012, pp. 629–632.

[40] C. Wang, Q. Wang, K. Ren, W. Lou, Privacy-preserving public auditing for data storage security in cloud computing, in: The 2010 IEEE Conference on Computer Communications(INFOCOM'10), IEEE, 2010, pp. 1–9.

[41] Y.S. Tan, R.K. Ko, P. Jagadpramana, C.H. Suen, M. Kirchberg, T.H. Lim, et al., Tracking of data leaving the cloud, in: 2012 IEEE 11th International Conference on Trust, Security and Privacy in Computing and Communications, IEEE, 2012, pp. 137–144.

[42] R.K. Ko, P. Jagadpramana, M. Mowbray, S. Pearson, M. Kirchberg, Q. Liang, et al., Trustcloud: a framework for accountability and trust in cloud computing, in: 2011 IEEE World Congress on Services, IEEE, 2011, pp. 584–588.

[43] R.K. Ko, M. Kirchberg, B.S. Lee, From system-centric to datacentric logging-accountability, trust & security in cloud computing, in: 2011 IEEE Defense Science Research Conference and Expo (DSR), IEEE, 2011, pp. 1–4.

[44] G.-H. Hwang, J.-Z. Peng, W.-S. Huang, A mutual nonrepudiation protocol for cloud storage with interchangeable accesses of a single account from multiple devices, in: 2013 12th IEEE International Conference on Trust, Security and Privacy in Computing and Communications (TrustCom), IEEE, 2013, pp. 439–446.

[45] Certicom Research, Elliptic Curve Cryptography, Standards for Efficient Cryptography (SEC) 1, 2000.

About the authors

Dr. Zhen Yang obtained his B.E. degree and Ph.D. degree in Electronic Information Science and Technology from Department of Electronic Engineering, Tsinghua University, Beijing, China, in 2012 and 2018. He is now a postdoctoral researcher in the Department of Electronic Engineering at Tsinghua University, China. His research interests are data security and privacy in cloud, and Internet of Things.

Miss. Yingying Chen is a graduate student, majoring in Master of Computer Science, at School of Engineering and Applied Science, University of Virginia, United States. She got her bachelor degree of Engineering in Internet of Things Engineering at International School, Beijing University of Posts and Telecommunications.

Dr. Yongfeng Huang is a professor at the Department of Electronic Engineering, Tsinghua University, Beijing, China. His research interests include Cloud Computing, multimedia network and next generation Internet. He has published 5 books and over 50 research papers on computer network and multimedia communication.

Dr. Xing Li is a professor of Department of Electronic Engineering, Tsinghua University, Beijing, China. His research interests include next generation Internet and computer network security.